Lecture Notes in Chemistry

Edited by G. Berthier, M. J. S. Dewar, H. Fischer
K. Fukui, H. Hartmann, H. H. Jaffé, J. Jortner
W. Kutzelnigg, K. Ruedenberg, E. Scrocco, W. Zeil

15

Alexandru T. Balaban
Adrian Chiriac
Ioan Motoc
Zeno Simon

Steric Fit in Quantitative Structure-Activity Relations

Springer-Verlag
Berlin Heidelberg New York 1980

Authors

Alexandru T. Balaban
Department of Organic Chemistry
Polytechnical Institute of Bucharest
Spl.Independentei 303
Bucharest 77206/Rumania

Adrian Chiriac
Department of Natural Sciences
University of Timisoara
Bd.V.Parvan 4
Timisoara 1900/Rumania

Ioan Motoc
Chemical Research Centre
Bd.Mihai Viteazu 24
Timisoara 1900/Rumania

Zeno Simon
Department of Biophysics
Institute of Medicine
Pta 23 August 2
Timisoara 1900/Rumania

ISBN 978-3-540-09755-6   ISBN 978-3-642-48316-5 (eBook)
DOI 10.1007/978-3-642-48316-5

Library of Congress Cataloging in Publication Data. Main entry under title: Steric fit in
quantitative structure-activity relations. (Lecture notes in chemistry ; 15) Includes
bibliographical references and index. 1. Structure-activity relationship (Pharmacology)
2. Biochemorphology. I. Balaban, Alexandru T. QP906.S75S73. 615'.7. 80-11453

STERIC FIT IN QUANTITATIVE STRUCTURE-ACTIVITY RELATIONS

A.T.BALABAN

A.CHIRIAC

I.MOTOC

Z.SIMON

CONTENTS AND CONTRIBUTORS

Addresses : A.T.Balaban: Department of Organic Chemistry, Polytech-
nical Institute of Bucharest, Spl.Independentei 303, Bucuresti 77206 ;
A.Chiriac, Department of Natural Sciences, University of Timisoara,
Bd.V.Parvan 4, Timisoara 1900; I.Motoc, Chemical Research Centre, Bd.
Mihai Viteazu 24, Timisoara 1900; Z.Simon, Department of Biophysics,
Institute of Medicine, Pta 23 August 2, Timisoara 1900; S.Holban,
R.Vancea and D.Ciubotariu, Polytechnical Institute of Timisoara,
Bd. 30 Decemvrie 6, Timisoara 1900, Romania.

Detailed Contents

1. INTRODUCTION

Although the importance of steric fit for receptor-effector interactions was recognized since Emil Fischer[1] proposed his "lock and key" theory, the whole area of steric properties is still in a very early stage of development.[2-4] We have a fairly good idea about electronic and hydrophobic parameters, but it is not easy to describe steric shapes of molecules without a large number of data. There are several cases of good QSAR's developed for rather large series of molecules without steric parameters — for example see papers by Hansch[5], or Franke[6], but the state of steric parameters is nevertheless one of the most important drawbacks, especially concerning the ability of encompassing, within a single QSAR, molecules of different shapes and stereoisomers. From today's steric parameters, one may mention the Taft[7] parameters E_S, which gave good results in organic chemistry, the rather cumbersome way of measuring shape differences of Amoore[8-10] and Allinger[11], and the L, B_1-B_4 parameters of Verloop[12].

The work described here consists of two types of approaches to the steric fit problem. The first approach consists of developing new parameters to describe different characteristics of the molecular shape (i.e., branching, bulkiness); this is done by means of topological indices. The second approach is based on minimal steric differences, a measure for steric misfit which depends not only on molecular shape, but also on the receptor and allows a guess of the shape of the receptor cavity. Brief reviews will be given on other steric parameters which will often be compared, in QSAR's, with parameters developed by the authors of these Lecture Notes. Electronic parameters and hydrophobicity-intermolecular force parameters, to be used together with steric parameters in QSAR's, are also briefly discussed. Other items include the metric introduced by minimal steric difference and computer programs developed in connection with our steric parameters.

A chapter reviews the use of topological indices in chemical, physico-chemical, and biological correlations. The philosophy of this approach rests on the translation of molecular structures, especially of the chain branching or of the general "topological shape", into a numerical index which can then lead to useful correlations and predictions. Among these topological indices, the centric index was introduced to correlate octane numbers of acyclic hydrocarbons, and the normalized and binormalized indices were introduced in order to reflect the overall shapes i.e., the nature of branching.[13]

The following abbreviations will be used for the steric para-

meters introduced by us : For the parameters based on topological indices[13], centric index, B; normalized centric index C; normalized quadratic index, Q; other abbreviations will be indicated in Chapter 3.

MSD — for the initial version of minimal steric differences
calculated simply as number of nonsuperposable (nonhydrogen) atoms
when the molecule is superimposed on the standard molecule, in a low
energy conformation allowing maximal superposition.[14-16]

MCD — for the minimal nonoverlappable volume when the molecule
is superimposed upon the standard molecule which is calculated by use
of a Monte Carlo technique.[17-19]

MTD — for the optimization technique which, starting from the
experimental activities of the correlated molecules, allows to obtain
the best probable shape of receptor cavity, receptor walls, and steric
irrelevant regions; this is a technique with a largely topologic character.[20-21]

The abbreviations MSD, MCD, and MTD will be used also for the
value of minimal steric differences calculated by the initial version,
by the Monte Carlo technique and, respectively, by the optimization
technique.

The experimental biological activities used in correlations
are almost always logarithms of reciprocal of active concentrations of
doses or of equilibrium constants. They will be denoted by A or A_i —
to mention explicitly the compound M_i to which they refer. The literature source for the experimental activities A used in correlations
is usually not indicated; it may be found in the bibliography of the
cited QSAR-paper. The calculated activities, resulted from correlational equations will be denoted by $\hat{A}$ or $\hat{A}_i$.

Anexes describe computing programs for MCD — by the Monte
Carlo technique and for MTD — by the optimization technique (the so-
called receptor mapping).

2. STERIC AND OTHER STRUCTURAL PARAMETERS FOR QSAR

Remembering that out of 4000-5000 synthesized compounds after
screening usually only one proves to be a useful therapeutic agent,the
efforts to develop theoretical methods for drug design are easily jus-
tifiable.[22] The generic term for these methods is Quantitative Structure Activity Relationships — QSAR.

The number of QSAR papers published during the 1962-1969 period[23] suggestively illustrates the interest for this domain : 2 papers in 1962/63, 11 in 64/65, 24 in 66/67 and more than 45 in 68/69.

About half of these publications comes from industrial research labo-
ratories.

The QSAR methodology uses a wide diversity of mathematical
techniques, the most frequently used ones being the following :
    1) cluster analysis [24-26]
    2) simplex algorithm [27,28]
    3) nonlinear models [29,30]
    4) discriminant analysis [31,32]
    5) pattern recognition [33-38]
    6) linear regression [39-41]

2.1. Correlational Equations and Predictor Variables

We will discuss in the present paper only linear regression [42-45] (uni- or multiparametric). We will therefore present briefly
the main ideas of this method.

If A_i is the biological response generated by the active compound (molecule) M_i, one considers $\hat{A}_i$ to be estimated by the linear expression (1) :

$$\hat{A}_i = a + \sum_{j=1}^{k} b_j \sigma_{ij} \; , \quad i = 1,2,\ldots, N, \; k \geq 1 \qquad (1)$$

σ_{ij} are denominated predictor variables or structural parameters. The
σ parameters give a numeric expression to topological, steric, and
electronic characteristics of M_i and/or physico-chemical properties of
biologically active molecules.

The regressional coefficients b_j are determined by the method
of least squares by condition (2) :

$$\text{MIN} \left(\sum_{i=1}^{N} (A_i - \hat{A}_i)^2 \right) \qquad (2)$$

Equation (1) may be used in one (or several) of the following purposes :
    i) pure description
    ii) prediction and estimation
    iii) extrapolation
    iv) estimation of parameters
    v) control
    vi) model building

The utilization of Equation (1) is conditioned by its statistical quality. Usually, regressional equations are characterized by correlation coefficient (r) and standard deviation (s). These two parameters do not allow a complete analysis of the quality of equations. It is necessary to use also other criteria such as the values of F (Fischer statistics), t (Student statistics), and EV (explained variance).

Another very important problem is multicolinearity of predictor variables (for a recent review, see Hotching[46]). By multicolinearity one understands the existence of functional relations f between some of the predictor variables :

$$\sigma_j = f(\sigma_i, \ldots, \sigma_{i+\ell}) \quad , \quad \ell = 0,1,\ldots \tag{3}$$

f being a linear function (or in principle any other type of function).

The main causes which produce multicolinearity[47] are :

a) an over-defined model

b) sampling techniques

c) physical constraints in the model or in the population.

In QSAR-work, multicolinearity is considered inexistent if the intercorrelation coefficients are low enough, $r^2(\sigma_i, \sigma_j) \lesssim 0.4$, an upper limit of 0.4 is considered as maximal for r^2 in case of chance correlation[48]. This test is necessary but not sufficient.

The model on which Eq.(1) is based requires also the following conditions to be satisfied :

1) The experimental A values must be measured with the same precision.

2) The values of the predictor variables, σ, must be determined with a sensibly higher precision than that of the dependent variable, A[49].

3) The regressional Equation (1) must be stable, i.e., small perturbations in the A-values, the increase or decrease of A values used to calibrate the regressional coefficients, should not modify strongly the equation.

The predictive power of Eq.(1) is conditioned, besides the above mentioned conditions, also by the size of the domain in the parameter space, explored by the parameters σ corresponding to the N compounds M_i.

Quantitative structure activity relations must be considered as semiempirical. Up till now, no completely satisfying theoretical justification has been given, although considerable efforts have been made in this direction.[50-52]

The interpretation of QSAR results and methodology within the framework of the traditional receptor concept[53,54] is relatively easy. The dynamic receptor concept (for a review, see Williams[55]) introduces considerable theoretical difficulties concerning the conformational aspects which are of primordial importance here. An attempt to deal with this problem for β-agonists is given by the fractional conformer effect of Richards[56]. In our opinion, the results given by QSAR refer only to the "excited" form of the receptor, i.e., to the receptor as it interacts with the effector.

In the following, this chapter gives a rather detailed account about steric parameters used in QSAR-studies, a brief review on electronic and intermolecular force parameters as well as a discussion on the use of indicator variables.

2.2. *Steric Parameters*

The steric parameters more frequently used in QSAR are analyzed in this paragraph. Of the less common steric parameters we mention the parachor[57,58], the Exner[59] molar volume (MV), the van der Waals volume (V) of Bondi[60]; for reviews on these parameters and their correlational power, see Lien[61], Verloop and Tripker[12], and Coburn and Solo[62].

Steric parameters, X, can be used in models of types :

$$\hat{A} = a + bX + \Sigma C_j \sigma_j \qquad (4)$$

$$\hat{A} = a + b_1 X + b_2 X^2 + \Sigma C_j \sigma_j \qquad (5)$$

$$\hat{A} = a + b \left| X_o - X \right| + \Sigma C_j \sigma_j \qquad (6)$$

with σ_j for nonsteric effects and X_o as optimum for X. Equation (5) allows in principle to determine the optimal shape and size of effector molecules for the biological response. Since for several steric parameters the detailed nature of steric interactions they express is not exactly known, one may recommend to use alternatively several types of steric constants, increasing thus the reliability for the observed correlations.

2.2.1. *Taft Type Steric Parameters* : E_S, E_S^c, E_S^o, E_S^e *and* ν

Following a proposal of Ingold[63], Taft[7] defined the steric substituent constant E_S by equation (7) :

$$\log(k_R / k_o) = \delta E_S \qquad (7)$$

using[46] the acid catalyzed hydrolysis of RCOOR' esters as standard reaction with $\delta \equiv 1.0$. The scale of E_S values was fixed by definition
with methyl (CH_3COOR) as origin ($E_{S,CH_3} \equiv 0$). Consequently bulkier
substituents R than CH_3 will have $E_{S,R} < 0$, retarding the hydrolysis
reaction, while less bulky substituents, $E_{S,R} > 0$. For a recent critical review on the work of Taft, see ref[64].

Several authors pointed out that E_S is not a purely steric
constant. This problem is interesting owing to the following considerations : let us suppose that E_S expresses steric effects (SE) with
the weight α and electronic effects (EE) with weight β. The type of
electronic effects is irrelevant in this context. Thus :

$$E_{S,R} = \alpha(SE)_R + \beta(EE)_R .$$

If R is electron accepting (by -I,-E effects) it tends to destabilize
the transition state of hydrolysis and with respect to Eq.(7), R will
appear more "voluminous" than it really is. In order to get rid of
this defficiency "corrected" E_S values were introduced :

a) E_S^c defined by Hancock[63-68] to eliminate hyperconjugative
effects of H atoms in α-position (N_H = number of such H atom) :

$$E_S^c = E_S + 0.306 \ (N_H - 3) \qquad (8)$$

b) E_S^o defined by Palm[69,70] to eliminate hyperconjugative effects of both H and C atoms (N_C = number of C-atoms in α) :

$$E_S^o = E_S + 0.33 \ (N_H - 3) + 0.13 \ N_C \qquad (9)$$

c) E_S^e defined by Unger[71,72] so as to eliminate both inductive
and conjugative electronic effects. Swain-Lupton[73] separation was used
and Equation (10) was obtained for CH_2X substituents :

$$E_{S,CH_2X}^e = E_{S,CH_2X} + 1.07 \ F_X + 1.05 \ R_X \qquad (10)$$

F_X and R_X are the "orthogonalised" quasi-inductive and quasi-resonant
Swain-Lupton constants.

Charton[74,75] studied the relation between van der Waals radii

$r_{V,X}$ and $E_{S,X}$ values and established a negligible contribution of σ_I and σ_R (inductive, resp. conjugative effects) for Eq.(11) :

$$E_{S,X} = a\sigma_{I,X} + b\sigma_{R,X} + cr_{V,X} + d \qquad (11)$$

For CH_2X type substituents Charton[74] obtained equation :

$$E_{S,X} = -0.412 \; r_{V,X} + 0.445 \; , \quad N=7, \quad r=0.915, \quad t=5.08 \qquad (12)$$

where N is the number of correlated compounds, r — the linear correlation coefficient, t — the Student test value, s — the mean standard deviation while for symmetric substituents (like NO_2, etc.) according to Kutter and Hansch[76] :

$$E_S = -1.839 \; r_{V,av} + 3.484; \quad N=6, \quad r=0.996, \quad s=0.132 \qquad (13)$$

$r_{V,av}$ represents average van der Waals substituent radii. For a critical review of Eqs.(12) and (13) see ref[72].

Conclusions generated by model (11) have led to definition of the steric parameter v_X (Charton[74,75]) corresponding to Eq.(14) :

$$v_X = r_{V,X} - r_{V,H} = r_{V,X} - 1.20 \qquad (14)$$

The v-values are determined by a Taft-type methodology.

Used in correlation of antifungal activity[77] (towards S.cerevisiae) of 13 salicylaldehyde derivatives (I), v yields Eq. (15)[78]:

$$A \equiv \log 1/c = 3.981 + 0.202(v_{R_1} + v_{R_3}) + 1.051(\sigma_{R_1} + \sigma_{R_2} + \sigma_{R_3}) \qquad (15)$$

I.

r=0.924, s=0.292, F=17.62, EV=0.825

R_1 = H, Cl, Br, I, CH_3

R_2 = H, Cl, Br, I, CH_3

R_3 = H, Cl

Table 1 illustrates Taft-type constants for 11 substituents.

Table 1. Taft-type steric substituent constants

Substituent	E_S	E_S^c	E_S^o	E_S^e	ν
CH_3	0.00	-1.24	-1.24	-1.24	0.52
C_2H_5	-0.07	-1.62	-1.51	-1.31	0.56
$n - C_6H_{13}$	-	-1.85	-1.69	-1.54	0.73
$n - C_3H_7$	-0.36	-1.91	-1.80	-1.60	0.68
$n - C_4H_9$	-0.39	-1.94	-1.83	-1.63	0.68
$n - C_5H_{11}$	-0.40	-1.95	-1.84	-1.64	0.68
$i - C_4H_9$	-0.93	-2.48	-2.37	-2.17	0.98
$(CH_2)_2Br$	-	-2.55	-2.43	-2.24	-
$(CH_2)_2I$	-	-2.57	-2.44	-2.26	0.93
$t - C_4H_9$	-1.54	-3.70	-3.38	-2.78	1.24
$CH_2C(CH_3)_3$	-	-3.29	-3.18	-2.98	1.34

E_S according to ref.[74]
E_S^c, E_S^o, E_S^e according to ref.[72]
ν according to ref.[71]

2.2.2 Amoore Type Steric Parameters

In an attempt of quantitative formulation for Ruzicka's[80] stereochemical theory of olfaction, Amoore[8] introduces the steric parameter $\bar{\Delta}$, a numeric expression for the similarity degree of shape and size of two compared molecules. The procedure of Amoore consists of[8-10] :

i) "front", "top" and "right"-silhouettes (i.e., projections on the corresponding planes) of the two molecules are obtained (photos of space-filling models for the molecules).

ii) pairs of silhouettes of the same type are superposed respecting superposition of weight centers and co-aliniation of main axis.

iii) on the thus superposed silhouettes, radii are traced with 10° spacings, from the weight center towards periphery. Absolute differences between corresponding radii in silhouette pairs (in $\mathring{A}$) are added to yield Δ.

iv) the average of the three Δ-indexes for the three types of silhouettes gives the steric parameter .

Calculations may use $\bar{\Delta}$ or the normalized $\bar{\Delta}_N = 1/(1+\Delta)$. Figure 1 illustrates the calculation of $\bar{\Delta}$ for a sphere inscribed within a cube. The advanced version[9] of the method determines $\bar{\Delta}$ by use of the PAPA machine (Probabilistic Automatic Pattern Analyser, a combination of photocamera, interface, and computer). The 36 radii used for hand-computation of $\bar{\Delta}$ are substituted by a reproducible collection of 4096 randomly selected straight lines. For applications of the method see ref[10,81].

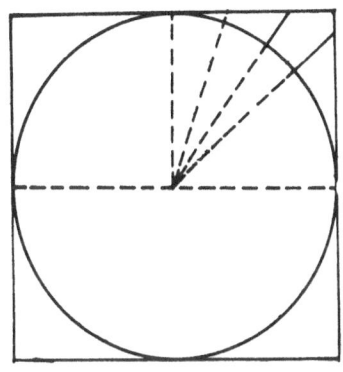

Figure 1. Calculation of Δ. Radii differences are marked as (—).

As an example of the correlative power of the $1/(1+\bar{\Delta})$ parameter, the dependence of odor similarity of N=30 organic compounds on shape similarity with the standard molecule of the ETH, CAM, and MOS-receptors was studied using values of ref[81], listed in Table 2. The following correlational equations resulted :

$$\hat{A}_1 = -5.486 + 12.834\ \bar{\Delta}_{N1}$$

$$r = 0.62,\quad s = 1.25,\quad F = 8.33,\quad CV = 0.36$$
(16)

$$\hat{A}_2 = -1.703 + 6.254\ \bar{\Delta}_{N2}$$

$$r = 0.61,\quad s = 0.48,\quad F = 7.88,\quad EV = 0.35$$
(17)

$$\hat{A}_3 = -0.240 + 2.848\ \bar{\Delta}_{N3}$$

$$r = 0.34,\quad s = 0.41,\quad F = 1.75,\quad EV = 0.08$$
(18)

The results are statistically highly significant (Eq.17 and 18) although not good, which may be justified by at least two considerations : (i) the precision of experimental data (odor similarities A obtained as average from a jury of 29 persons) may not be very high, and (ii) silhouette superpositions (odorant vs.standard) are performed without an objective criterion.

For the test of the correlational value of the $1/(1+\bar{\Delta})$ parameter one may recommend the use of a less difficult series than the odorants, especially a series for which previous QSAR-studies indicate the existence of steric interactions.

Table 2. Data[a] used for derivation of Eqs.(16)-(18)

No.	Compound	ETH[b]		CAM[c]		MUS[d]	
		A_1	$\overline{\Delta}_{N1}$	A_2	$\overline{\Delta}_{N2}$	A_3	$\overline{\Delta}_{N3}$
1.	Acetone	4.42	0.687	1.04	0.460	0.63	0.378
2.	Acetophenone	0.71	0.584	2.00	0.567	2.08	0.461
3.	Acetylene tetrachloride	4.32	0.647	1.05	0.573	0.57	0.464
4.	Anisic aldehyde	0.77	0.579	2.53	0.538	1.73	0.506
5.	Anisole	2.65	0.630	1.44	0.554	0.91	0.468
6.	Benzaldehyde	1.27	0.643	1.73	0.513	1.14	0.433
7.	Benzene	3.62	0.639	1.10	0.493	0.70	0.403
8.	Benzonitrile	1.14	0.640	2.54	0.506	1.32	0.431
9.	Benzophenone	1.05	0.505	1.42	0.539	1.18	0.501
10.	Benzyl acetate	1.65	0.623	1.69	0.567	1.68	0.469
11.	n-Butyl propionate	1.32	0.599	1.47	0.460	1.17	0.478
12.	Chlorobenzene	2.16	0.668	1.47	0.518	1.10	0.430
13.	Chloroform	6.69	0.645	0.60	0.488	0.24	0.388
14.	Cyclohexane	3.66	0.609	1.78	0.593	0.62	0.448
15.	Cyclohexanol	1.51	0.618	2.25	0.617	0.74	0.469
16.	Cyclohexanone	1.91	0.627	2.06	0.614	1.77	0.465
17.	Cyclooctanone	1.55	0.545	3.28	0.658	0.29	0.520
18.	Cyclopentanone	2.33	0.698	2.05	0.339	0.87	0.427
19.	Cyclopentyl acetate	0.91	0.580	1.75	0.599	1.68	0.509
20.	Di-n-Butyl ether	1.29	0.582	0.87	0.435	1.14	0.471
21.	Cis-1,2-Dichloroethene	5.45	0.672	0.71	0.455	1.67	0.374
22.	Trans-1,2-Dichloroethene	3.41	0.793	1.26	0.450	0.74	0.372
23.	Diethyl ether	5.12	0.725	1.86	0.450	0.69	0.412
24.	Diethyl sulfate	1.86	0.531	1.90	0.551	1.32	0.517
25.	Dimethylbenzylcarbinol	1.02	0.504	1.48	0.633	1.25	0.535
26.	Dimethylbenzylcarbinyl-acetate	1.07	0.457	2.30	0.607	1.14	0.565
27.	2,2-Dimethylpropyl acetate	1.34	0.546	2.35	0.582	1.54	0.478
28.	Diphenyl ether	1.14	0.503	1.07	0.566	1.04	0.485
29.	Ethyl acetate	3.48	0.698	1.23	0.503	0.67	0.423
30.	Eugenol	1.28	0.498	1.90	0.535	1.55	0.557

a) A's are odor similarities, Δ_{NI} the normalized parameters $1/(1+\Delta)$, both with respect to the corresponding standard
b) ETH - etheral receptor, standard 1,2-dichloroethane
c) CAM - camphoraceus receptor, standard 1,8-cinneol
d) MUS - musky receptor, standard 15-hydroxypentadecanoic acid.

The methods of topographic analysis of molecules of the John Hopkins University group, rely on Amoore's methodology. Thus, the "4-π comparator" method (Wilson[83]) uses CPK molecular models, Stereoprojections (photographs realized with telescope lenses) along six orthogonal directions, (corresponding to the faces of the cube containing the respective molecule) for the studied compounds reveal shape and volume differences among the molecules of the series. The advanced version[84] generates the six projections by means of a computer with a corresponding plotter.

2.2.3 *Molecular Refractivity and van der Waals Volume*

Molecular refractivity, MR, was introduced as steric parameter in QSAR studies by the Pomona College group[85-90], instead, or together with the Taft parameter, E_S.

MR is defined by the Lorentz-Lorenz Equation (19) :

$$MR = \frac{n^2-1}{n^2+2} \cdot \frac{MW}{d} \tag{19}$$

where n is the refractive index, MW is the molecular weight and d is the density. MR is an additive parameter. For an extensive list of MR values, see refs.[72,88,89]

Relation (19) indicates that MR contains both a dielectric component measured by London dispersion forces, as well as one of steric repulsion. Thus MR is not a purely steric constant and one must pay attention to multicolinearity.[90] Kaufman[91] points out an exponential correlation between the partition coefficient P and MR (i.e., linear correlation between π or f and MR).

Since from a sterical view point the MR is considered to be a measure for the van der Waals volume one should be careful in testing this assumption. Van der Waals volumes can be calculated by numeric integration (using the Monte-Carlo technique) of the van der Waals envelope of the corresponding substituent, similar to the method described in Chapter 6. Numeric calculations have been performed by means of the NOVA/1 program, developed by Motoc and detailed description of results will be given elsewhere.[92] Here only a brief account of part of the results is given. Standard values for bond lenghts and angles are used, and the van der Waals radii are listed in Table 3.

Table 4 gives MR values and van derWaals volumes (MV), for 20 substituents. With them the following regressional equations are obtained.[95]

$$MR = -2.510 + 0.380 \ WV_A \tag{20}$$

$$N = 20, \quad r=0.943, \quad s=1.578, \quad F=68.6, \quad EV=0.884$$

$$MR = -2.678 + 0.337 \ WV_B \tag{21}$$

$$N = 20, \quad r=0.949, \quad s=1.501, \quad F=76.8, \quad EV=0.915$$

Table 3. Atomic van der Waals radii (Å)

Atom	WV_A[93]	WV_B[94]
H	1.24	1.20
C	1.54	1.57
F	1.18	1.35
Cl	1.63	1.80
Br	2.02	1.95
I	1.02	2.15
N	1.36	1.50
O	1.27	1.40

The two equations indicate MR as a good measure for van der Waals substituent volumes. Equations (20) and (21) are of practically the same statistical quality and express the same dependence between MR and WV, in spite of the sometimes rather large differences between van der Waals radii used in calculations.

As an example, MR and WV are used in a comparative study of phenylsuccinamate (II) interaction with the antibody against 3-azopiridine. Part of the results of Hansch and Moser[96] are used for comparison. Experimental data, ΔF_{rel} and parameters σ and π are listed in Table 5; the MR and WV values of Table 4 were employed.

II.
$$\text{R} \quad \bigcirc \text{—NHCOCH}_2\text{CH}_2\text{COO}^-$$

The following equations were obtained

$$\Delta F^o_{rel} = -2.380+0.590\sigma-0.237\pi+0.117 \ MR^m+0.173 \ MR^p \tag{22}$$

$$N=20, \quad r=0.948, \quad s=0.242, \quad r(\sigma)=0.412, \quad r(\pi)= -0.271$$

$$r(MR^m)=0.271, \quad r(MR^p)=0.648, \quad F=24.96, \quad EV=0.872$$

Table 4. MR and WV values

No.	Substituent R	$MR^{a)}$	$WV^{b)}$	
			WV_A	WV_B
1.	CH_3	5.7	24.86	23.08
2.	CH_2Cl	10.5	38.16	41.71
3.	CH_2Br	13.4	41.07	46.40
4.	CH_2I	18.6	51.14	57.12
5.	$CH F_2$	5.2	26.76	31.41
6.	CF_3	5.0	27.33	34.55
7.	$CHCl_2$	15.3	46.19	55.97
8.	F	0.8	6.85	10.29
9.	Cl	5.8	18.29	24.44
10.	Br	8.7	24.03	31.01
11.	H	1.0	7.91	7.22
12.	I	14.0	34.39	41.52
13.	NO_2	6.7	20.84	26.69
14.	$CO-CH_3$	11.2	36.15	37.87
15.	NH_2	4.4	18.29	19.40
16.	CN	5.5	20.43	23.38
17.	CH_2OH	7.2	30.71	31.12
18.	COOH	6.9	28.17	31.38
19.	OH	2.6	13.05	14.78
20.	COO^-	6.1	24.13	28.42

A) Calculated with van der Waals radii of ref.[93].
B) Calculated with van der Waals radii of ref.[94].
a) MR figures from ref.[72].
b) Estimated error for WV calculations about 0.5%.

$$- \Delta F^o_{rel} = -2.886 + 0.537\sigma - 0.207\pi + 0.041\ WV^m_B + 0.060\ WV^p_B \tag{23}$$

$$N = 20, \quad r = 0.941, \quad s = 0.258; \quad r(\sigma) = 0.412, \quad r(\pi) = -0.271,$$

$$r(WV^m) = 0.245; \quad r(WV^p) = 0.643; \quad F = 21.61; \quad EV = 0.855$$

Table 5. The values used in the derivation of Equations (22) and (23)

No	R	$-\Delta F^o_{rel}$	σ	π	$- F^o_{rel}$ [a]	$- F^o_{rel}$ [b]
1.	H	-1.90	0.00	0.00	-2.09	-2.15
2.	$2-CH_3$	-2.75	-0.17	0.56	-2.31	-2.36
3.	$3-CH_3$	-1.72	-0.07	0.56	-1.71	-1.65
4.	$4-CH_3$	-1.73	-0.17	0.56	-1.51	-1.40
5.	$2-Cl$	-1.96	0.23	0.71	-2.13	-2.17
6.	$3-Cl$	-1.59	0.37	0.71	-1.47	-1.39
7.	$4-Cl$	-0.81	0.23	0.71	-1.27	-1.14
8.	$2,4-di-Cl$	-1.71	0.46	1.42	-1.32	-1.24
9.	$2,5-di-Cl$	-1.33	0.60	1.42	-1.52	-1.47
10.	$2-Br$	-1.94	0.23	0.86	-2.17	-2.21
11.	$3-Br$	-1.43	0.39	0.86	-1.16	-1.14
12.	$4-Br$	-0.74	0.23	0.86	-0.81	-0.77
13.	$2-I$	-0.02	0.18	1.12	-2.26	-2.29
14.	$3-I$	-0.43	0.35	1.12	-0.64	-0.78
15.	$4-I$	-0.18	0.18	1.12	-0.03	-0.22
16.	$2-NO_2$	-1.74	0.78	-0.28	-1.61	-1.67
17.	$3-NO_2$	-1.17	0.71	-0.28	-0.90	-0.91
18.	$4-NO_2$	-0.61	0.78	-0.28	-0.52	-0.50
19.	3-Acetyl	-0.48	0.38	-0.55	-0.56	-0.57
20.	4-Acetyl	0.32	0.50	-0.55	0.06	0.08

a) Calculated using equation (22)
b) Calculated using equation (23)

The two equations confirm the importance of dispersive interactions between the haptene and the receptor.

As a conclusion, MR expresses with reasonable accuracy the van der Waals substituent volume. Taking, nevertheless, into account the clear physical significance and the possibility to be determined by computation, one may recommend to use the new parameter WV, instead of MR.[*]

---

[*] The FORTRAN-IV program for computing WV values can be obtained from dr. I.Motoc.

2.2.4. Verloop-Hoogenstraaten-Tipker Steric Parameters

Verloop, Hoogenstraaten, and Tipker[4,12] introduced the STERIMOL substituent parameters to characterize shape and size of substituents. This is a set of five parameters B_1, B_2, B_3, B_4, L . B_1-B_4 measure the thickness of the substituent along four directions perpendicular on its main axis, L-its length. These parameters are computed with the FORTRAN STERIMOL program[97], which uses a linear code for description of structural formula so that it can be applied without further difficulties.

The B_1-B_4,L parameters have at least two important advantages in comparison to other steric parameters : (i) they describe more adequately nonsymmetrical substituents and emphasize the directional (vectorial) character of steric effects, extending thus the possibilities of QSAR analysis, (ii) they offer more detailed information about the actual conformation of the receptor when it is attacked by the effector. Nevertheless, the fact that it introduces at once five predictor variables in the regressional analysis requires relatively large series of data in order to obtain reliable results.

Table 6 lists STERIMOL-constants for 11 substituents (compare with Table 1) according to ref[12].

Table 6. Steric STERIMOL-constants

Substituent	B_1	B_2	B_3	B_4	L
CH_3	1.52	1.90	1.90	2.40	3.00
C_2H_5	1.52	1.90	1.90	2.97	4.11
$n\text{-}C_6H_{13}$	1.52	1.90	1.90	5.87	8.22
$n\text{-}C_3H_7$	1.52	1.90	1.90	3.49	5.05
$n\text{-}C_4H_9$	1.52	1.90	1.90	4.42	6.17
$n\text{-}C_5H_{11}$	1.52	1.90	1.90	4.94	7.11
$i\text{-}C_4H_9$	1.52	1.90	3.16	4.21	5.05
$(CH_2)_2Br$	1.52	1.95	1.95	3.40	5.87
$(CH_2)_2J$	1.52	2.15	2.15	3.60	6.28
$t\text{-}C_4H_9$	2.59	2.86	2.86	2.97	4.11
$CH_2C(CH_3)_2C_2H_5$	1.52	3.16	3.16	4.42	6.17

As an illustration of the correlational power of STERIMOL parameters we present correlational studies concerning in vitro microsomal p-hydroxylation of substituted anilines of type III. For A=log k_r (relative rate constant of hydroxylation reactions), Hansch[98] obtains Eq.(24), excluding m-OCH_3 of the analysis :

III.

$$\hat{A}=0.27 \log P+0.36E_S + 1.03 \tag{24}$$

$$N=11, \quad r=0.982, \quad s=0.082, \quad F=120.5$$

If m=OCH_3 is included the correlation deteriorates :

$$\hat{A}=0.45 \log P + 0.31 E_S + 0.66 \tag{25}$$

$$N=12, \quad r=0.856, \quad s=0.288, \quad F=12.38$$

With STERIMOL parameters, following equations are obtained[12]

$$\hat{A} =0.51 \log P - 0.48 B_1 + 1.39 \tag{26}$$

$$N=12, \quad r=0.818, \quad s=0.320, \quad F=9.12$$

$$\hat{A} =0.44 \log P - 0.33 L + 1.81 \tag{27}$$

$$N=12, \quad r=0.903, \quad s=0.240, \quad F=19.9$$

$$\hat{A} = \log P - 0.54 B_4 + 1.96 \tag{28}$$

$$N=12, \quad r=0.969, \quad s=0.138, \quad F=69.5$$

Equation (28) demonstrates that exclusion of m-OCH_3 cannot be justified phenomenologically (as tried in ref.[98]) but by the fact that E_S does not give, in this case, an adequate description of the steric characteristics of the substituent.

For other applications of STERIMOL-parameters cf. refs.[12,99].

The three versions of the minimal steric difference method for evaluation of steric effects are described in Chapters 4,5 and 6. Topological indices derived in the framework of chemical graph theory are discussed in Chapter 3. Good results obtained in the study of octane numbers for alkanes justify the attentive investigation of the possibilities of topological indices for QSAR-studies. Up to now, only the Randić index was studied in this respect (see refs. quoted in Chap.3).

2.3 Indicator Variables

Indicator variables were introduced in QSAR studies by Bruice et al.[100] and their use was substantiated by Free and Wilson[101] (de novo constants). Indicator variables were defined by relation

$$I = \delta_{ik} = \{0 \text{ or } 1\} \tag{28}$$

as the term (molecule) "i" of the series of compounds is substituted (with a certain substituent) in a position "k" (I=1) or is not (I=0). Nowadays the significance of the indicator variable I was extended to express the existence, in the structure of the molecule under consideration, of a given structural characteristics. The value range of I is a finite subset of the natural numbers.

As an illustrative application we describe here a study of dihydrofolate reductase by substituted pyrimidines of type IV[95] :

IV.

Following the study of 28 "de novo" indicator variables in several linear and nonlinear models, Hansch et al.[102] obtained the following best regressional equation (A ≡ log I/C refers to reversible inhibition) :

$$\hat{A}=0.365 \ I_1-1.013 \ I_8-0.784 \ I_9+0.419 \ I_{13}-0.220 \ I_{15}+0.513 \ I_{20} +$$
$$+0.674 \ I_4I_8+7.174 \ , \quad N=105, \quad r=0.903, \quad s=0.229 \tag{29}$$

Table 7. Indicator variables for Eq. (29)

I_j	Characteristic
I_1	$\omega = CH_2$
I_4	$y = 3-CH_3$
I_8	BRD=4-NHCONH-
I_9	BRD=4-NHCO-
I_{13}	BRD=4-NHSO$_2$-
I_{15}	Z = 4-SO$_2$F
I_{20}	indicator variable for the enzyme

The indicator variables I_j of Eq.(29) are I_j=1 if following structural characteristics exist (Table 7) in the alternative case I_j=0.

The nonlinear term I_4I_8 of Eq.(29) has a special significance, describing the contribution of cooperational interaction between

3-CH$_3$ and 4-NHCONH. The QSAR study revealed that 3-CH$_3$ gives a significant contribution to activity only in the presence of the BRD = 4-NHCONH-bridge.

De novo variables can be used also in hybrid models, with both continuous (σ, π, MR etc.) and discrete predictor variables. A good example is the interaction of type V and VI-compounds with papaine[103].

V :   (benzene ring with X substituent)— OCOCH$_2$NHSO$_2$CH$_3$;   VI :   (benzene ring with X substituent)—OCOCH$_2$NHCOC$_6$H$_5$

The substituent X is characterized by σ-Hammett and MR-constants and one obtains Equation (30) :

$$\hat{A} = 0.57 \text{ MR} + 0.56\sigma - 1.92 \text{ I} + 3.74$$
$$N=20, \quad r=0.990, \quad s=0.148$$

(30)

where   A = log 1/K$_m$, with K$_m$ - the Michaelis binding constant, I is a de novo variable with I=1 for type V structures and I=0 for type VI structures. Other examples see refs.[104,105]

Indicator variables with extended significance referring to intermolecular force types in receptor-oligopeptide interactions, are described in paragraph 2.5 .

The use of indicator variables in hybrid models and of indicator variables with extended significance has the drawback of defining large equivalence classes, rather difficult to justify. Certainly, there are several cases where indicator variables are the only possible alternative, especially for series of molecules differing by large substituents with complicated structure.

2.4 Electronic Parameters-Synopsis

Electronic substituent parameters used in QSAR studies were obtained by means of the $\rho\sigma$ formalism and by the formalism of quantum chemistry. The first category of parameters include :

σ - Hammett constant [106]

σ^+ - constant for electron-demanding reactions [107]

σ^- - constant for electron-releasing reactions [108]

σ^n - unexalted Hammett constant [109]

σ^0 - unenhanced Hammett constant [110]

σ^x - aliphatic polar constant [7]

σ' - aliphatic polar constant [111]

σ_I, σ_R - inductive and resonance constants [112]

σ_o - polar constant for ortho-substituent (general) [112]

σ_o^* - polar constant for ortho-substituent relative to methyl [113]

F,R - field and resonance constants [73]

n - number of bonds allowing for hyperconjugation [114]

σ^ϕ - Hammett constants for substituents with P atom [115]

The second category of parameters include[116-120] :

Energy of the highest occupied molecular-orbital (E_{HOMO}) - Measure of electron- donating ability of molecule

Energy of the lowest empty unoccupied molecular orbital (E_{LEMO}) or (E_{LUMO}) - Measure of electron accepting ability of molecule

Electron density (Q_r) - Electronic charge associated with atom r,

Frontier electron density (f_r^E) - Electric charge of electrons in the HOMO associated with atom r.

Nucleophilic superdelocalizability (S_r^N) - measure of electron-donating ability of atom r

Electronic superdelocalizability (S_r^E) - measure of electron-accepting ability of atom r

The composition of σ constants, inductive vs. conjugative component :

$$\sigma = \lambda\sigma_I + \delta\sigma_R \qquad (31)$$

expressed by the percentage of conjugative component, $\varepsilon = 100\ \delta/(\lambda+\delta)$, for different σ constants is[120] :

σ_I:0%; σ_m:25%; σ_p^o:40%; σ_p:50%; σ_p^+:62%; σ_p^-:60%; σ_R:100%

An interesting example for the use of quantum chemical indices is the work of Peradejordi et al.[126]. For two reviews on this item, cf.refs.[117] and [118]. Concerning the use of MO-techniques for calculating quantum indices in QSAR-studies, Charton[120] recommends advanced methods such as PPP and EHT.

2.5 Intermolecular Force Parameters

Hydrophobic interactions are of uttermost importance for in-teractions between biomolecules[121-128]. The quantitative estimation of hydrophobic character is based on the use of partition coefficients between hydrophobic solvent (n-octanol) and water. Two semi-empirical methods where developed for this purpose :

    1) The method of Hansch et al.[129-132] uses equation :

$$\log(P_{SX}/P_{SH}) = \rho\pi_X \tag{32}$$

to determine hydrophobic substituent-X constants. P_{SX} and P_{SH} represent partition coefficients for compounds SX and SH. Eq.(32) was standardized by using the octanol / water solvent pair as standard system, for which $\rho = 1$[133]. For polysubstituted compounds $SX_1 \ldots X_N$ Eq.(32) becomes :

$$\log(P_{SX_1 \ldots X_N}/P_{SH}) = \sum_{i=1}^{} \pi_{X_i} \tag{33}$$

Corrections for different interactions, branching etc., are also used [132].

    Equations of the type (32) and (33) proved of high value for QSAR studies. Some drawbacks (for example one obtains $\pi_H=0$, $\pi_{CH_3} = \pi_{CH_2}$) are discussed in ref.[134] and [64], while a different point of view on the π-system is given in ref.[135].

    2) The method of Rekker and Nyss[136] eliminates the drawbacks discussed above. Hydrophobic fragmental constants f_X are defined according to :

$$\log P_{SX} = f_S + f_X \tag{34}$$

or generally for a compound $SX_1 \ldots X_N$:

$$\log P_{SX_1 \ldots X_n} = f_S + \sum_{i=1}^{n} a_i f_i + \sum_{k} p_k \tag{35}$$

where a_i is the number of times fragment X_i appears in the molecule $SX_1 \ldots X_N$, p_k are corrections for proximity effects which must be sometimes introduced and P are octanol/water partition coefficients. The method of Rekker and Nyss gives better results in calculating molecular hydrophobicity by additive principles and f_X-values may be recommended to be used instead of π_X-values.[137]

    As defined by the use of octanol : water partition coefficients, hydrophobicity generally gives a good idea of what happens when by effector-receptor combination, the surrounding water is replaced by the usually much less polar partner. Juxtaposition of groups so as to allow an optimal formation of hydrogen bonds, electrostatic interactions etc., is not directly taken into account by hydrophobicity. In order to describe directly such interactions, intermolecular force parameters were introduced (related especially to amino-acidic side chains)[15,138-140] :

- Hydrogen bond-forming groups, proton donors, HBD or acceptors, HBA (Alternatively, HB with ± 1, respectively).

- Electrically charged groups at pH 7, EC, with elementary charges +1 or -1, usually.

- Aromatic groups AR. van der Waals interactions are proportional to the polarizability product between the two partners and an aromatic ring should interact more strongly with another aromatic cycle (parallel and superposed) than with an aliphatic group.

- Charge transfer complex forming groups, electron donors, CTD, or acceptors, CTA.

- Groups with high dipole moments, μ, not hydrogen bond forming.

Some examples of QSAR studies with these types of intermolecular force parameters are given in paragraph 4.2.

Attempts to quantify hydrogen bond forming ability were made by Seiler[141] and Moriguchi et al.[142]. Seiler[141] introduced the parameter I_H defined as :

$$I_H = \log P_{octanol} - \log P_{cyclohexane} + b \qquad (36)$$

$P_{octanol}$ and $P_{cyclohexane}$ are octanole : water and, respectively, cyclohexane : water partition coefficients.

Although insufficiently developed, new parameters to quantify hydrogen bonding ability are certainly worth while to develop.

Hetnarsky and O'Brien[143] proposed a charge transfer parameter for QSAR, C_T, defined as :

$$\lg [K_X / K_H] = C_T \qquad (37)$$

where K_X is the equilibrium constant for charge transfer complex formation with tetracyanoethylene while K_H refers to the unsubstituted compound. Thus K_X refers to a substituent in an aromatic cycle (or unsaturated system).

Finally, one may mention that molecular refraction, MR, may be considered also to reflect van der Waals interactions (proportional to the product of polarizabilities) and that the parameters for different types of intermolecular force forming groups may be considered also as indicator variables.

3. TOPOLOGICAL INDICES

The constitutional or structural formula of a chemical compound may be viewed as a molecular graph, where the vertices represent atoms and the edges represent the covalent bonds. Graph theory[144,145] is a well-established branch of combinatorial mathematics and may be applied to chemical problems.[146,147] Among these problems, the characterization of the bonding topology and, if possible, of the stereochemistry of molecular graphs is a prominent one.

For chemical documentation purposes, each molecular graph should have a unique name or code which should allow indexing and retrieval. This necessity has been partly fulfilled by the traditional chemical nomenclature and by the development of linear notation systems. Both approaches are in use and are able to discriminate among isomeric structures, but rely heavily on conventions. Chemical names may be indexed alphabetically within a set of isomeric systems, which in turn may be ordered according to the molecular formula; this fact makes chemistry the best-documented natural science. Linear notation systems such as the Dyson(IUPAC) system,[148] or the more widely used Wiswesser Line Notation[149] may be implemented on computers making literature search for chemical structures or substructures rapid and reliable. In addition, topological codes were also developed but so far mostly for acyclic molecular graphs, though these are less numerous than graphs containing cycles. Since these topological codes need fewer conventions, they are easier to master, to apply and to implement on computers; in addition, they may also be used for enumeration (not only for codification) of isomers. Such is the DENDRAL program developed by Lederberg et al.[150] or the more limited topological code proposed by Read.[151] These two codes are based on the existence of a unique center (one vertex) or bicenter (two adjacent vertices, i.e., a pair of vertices linked by one edge) in any acyclic graph, as demonstrated in graph theory.[144]

For purposes of correlations between structure and properties, a different approach was used, namely, the bonding topology of each molecular graph is converted into an expression which may be a matrix, a polynomial, a sequence of numbers, or a numerical index. Numerical indices developed for this purpose are called topological indices.

Two matrices characterize uniquely molecular graphs : the adjacency matrix $\underline{\underline{A}}$ is most commonly used, and consists of entries $A_{ij} = A_{ji}$ which are 1 if vertices i and j are adjacent (i.e., linked by an edge), and 0 otherwise; the second matrix, also symmetrical, has as entries $D_{ij} = D_{ji} = d_{ij}$, where d_{ij} indicates the number of edges in the shortest path between vertices i and j. Since d_{ij} is called the distance between vertices i and j, the latter matrix is called the distance matrix $\underline{\underline{D}}$.

It was shown [152] that when the molecular graph represents the carbon skeleton of a conjugated hydrocarbon, the eigenvalues of the adjacency matrix are identical to those of the Hückel matrix, so that there exists a complete equivalence between the two treatments: graph-theoretical, and quantum-chemical in the Hückel M.O. approximation.

A long search was made to find a reduced form of the adjacency matrix, more easy to handle than the matrix, and yet able to characterize uniquely the molecular graph. Spialter asserted that the characteristic polynomial of the adjacency matrix (or the eigenvalues of the matrix A, i.e., the algebraic roots of the characteristic polynomial, which are called the spectrum of the graph) could function as such a reduced form.[153] The characteristic polynomial may be defined as

$$\det |\underline{\underline{A}} - x \; \underline{\underline{E}}| = 0$$

where $\underline{\underline{E}}$ is the unit matrix. This means that one replaces the zeroes of the main diagonal in the adjacency matrix by x's and that the resulting determinant is converted into polynomial form and equated to zero. The n roots of this characteristic polynomial are all real ; n is the number of vertices in the graph, or of carbon atoms in the compound. Spialter's assertion was contested by Balaban and Harary.[154] Japanese authors[155] argued that Spialter's assertion might be true if the molecular graph was not the hydrogen-depleted graph of the carbon skeleton in hydrocarbons, but included also the hydrogen atoms. However this was again found to be not true.[156] Therefore no simple reduced form of the adjacency matrix has yet been found, which could characterize uniquely the topology of molecules.

Randić observed[157] that if one rearranges the rows and columns of the adjacency matrix (these rearrangements correspond to different

numberings of the same molecular graph), and if one reads sequentially
the entries, a binary number results for each rearrangement. If one
chooses the minimal binary number, this number (which may be converted
into decimal notation for convenience) represents the topology of the
given molecule. At the same time, this number corresponds to a non-
conventional numbering of vertices in the molecular graph. Randić con-
jectured[158] that this procedure might lead to an index uniquely associated to the topology of the molecule, and despite counter-arguments
[159] this conjecture may be true.

Topological indices[147,160] need not, however, be unique. They
should merely reflect, in a more or less discriminating manner, the
molecular topology, and should allow correlations with the structure.

3.1 Enumeration of Topological Indices

In the following, the topological indices described so far in
the literature will be enumerated in roughly chronological order. They
will be exemplified by calculating in each case the values of the index for two molecules : (i) isooctane or 2,2,4-trimethylpentane, whose
carbon skeleton (hydrogen-depleted graph) is I, and (ii) n-heptane, II.
Under each structure we display the corresponding adjacency matrix,
and below the corresponding distance matrix. Finally, we enumerate the
distribution of distances as apparent from the distance matrix.

The first topological indices have been proposed by Wiener.[161]
In order to correlate boiling points and other properties of alkanes
(acyclic saturated hydrocarbons), he introduced two numbers : the path
number w (or the Wiener index, as Platt[162] called it), which is the
sum of the number of bonds between all pairs of vertices, and the
polarity number p, which is the number of pairs of vertices separated
by three edges. The number p indicates, after Wiener, the number of
steric pairs. For graph I, w = 66 and p = 5; for II, w = 56 and p=4.

More recently, Rouvray and Crafford[163] simplified the finding
of index w by devising a new index I which is the sum of all entries
in the distance matrix $\underline{D}$. Evidently, $I = 2w$.

Indices w and I reflect the branching, hence the topology of
the molecule, but index p does not. The last index, p, was used as an
extra factor for correcting correlations between w and boiling points,
or other properties of hydrocarbons, as will be shown later in this
review.

Structure I

Structure II

	I	II
Adjacency matrix	$\begin{Vmatrix} 0&0&0&1&0&0&0&0 \\ 0&0&0&1&0&0&0&0 \\ 0&0&0&1&0&0&0&0 \\ 1&1&1&0&1&0&0&0 \\ 0&0&0&1&0&1&0&0 \\ 0&0&0&0&1&0&1&1 \\ 0&0&0&0&0&1&0&0 \\ 0&0&0&0&0&1&0&0 \end{Vmatrix}$	$\begin{Vmatrix} 0&1&0&0&0&0&1 \\ 1&0&1&0&0&0&0 \\ 0&1&0&1&0&0&0 \\ 0&0&1&0&1&0&0 \\ 0&0&0&1&0&1&0 \\ 0&0&0&0&1&0&1 \\ 1&0&0&0&0&1&0 \end{Vmatrix}$
Distance matrix	$\begin{Vmatrix} 0&2&2&1&2&3&4&4 \\ 2&0&2&1&2&3&4&4 \\ 2&2&0&1&2&3&4&4 \\ 1&1&1&0&1&2&3&3 \\ 2&2&2&1&0&1&2&2 \\ 3&3&3&2&1&0&1&1 \\ 4&4&4&3&2&1&0&2 \\ 4&4&4&3&2&1&2&0 \end{Vmatrix}$	$\begin{Vmatrix} 0&1&2&3&4&5&6 \\ 1&0&1&2&3&4&5 \\ 2&1&0&1&2&3&4 \\ 3&2&1&0&1&2&3 \\ 4&3&2&1&0&1&2 \\ 5&4&3&2&1&0&1 \\ 6&5&4&3&2&1&0 \end{Vmatrix}$

Distance i		
No. of distances g_i	1 2 3 4	1 2 3 4 5 6
	7 10 5 6	6 5 4 3 2 1

If the number of pairs of vertices whose distance i is denoted by g_i, then the Wiener index w may be written

$$w = \sum_i i g_i \; .$$

Altenburg[164] modified the Wiener index by expressing it as a sum of terms depending on an indexed variable a_i. His expression is

$$\sum_i a_i g_i \; .$$

Both Wiener's index and Altenburg's expression were initially devised for acyclic graphs. Later, it was shown[165a] that they may be extended to cyclic graphs.

Platt[166] used the two Wiener numbers in many correlations, and introduced a third index, f (first-neighbor sum). This index is calculated by determining for each edge the number of adjacent edges, and then by summing these numbers for all edges. We present below examples I and II.

I (f = 20)   II (f = 10)

As indicated above in the case of two related indices, w and I, another index N_2 introduced by Gordon and Scantlebury[167] is related to a previously described index by a similar relation : $f = 2N_2$. Gordon's index N_2, defined as "the second moment of distributions for the number of units with degree of substitution i in the j-th n-isomer" may be calculated simply as the number of distinct ways a C-C-C fragment may be superimposed on the hydrogen-depleted graph. The first moment of distributions is 2(n-1). For an alternative definition of N_2 see below under quadratic indices. Indices N_2 for I and II result as follows.

I (N_2 = 10)   II (N_2 = 5)

Hosoya [165,168] defined for any graph G a topological index Z as

$$Z = \sum_k p(G,k)$$

where p(G,k) is the number of ways in which k from all q edges of graph G may be chosen so that no two of them are adjacent. For trees (acyclic graphs), an equivalent definition of Z is as the sum of the absolute values of coefficients in the characteristic polynomial :

$$\sum_{k=0}^{n} (-1)^k p(G,k) x^{n-2k} = (-1)^n \det| \underline{A} - x \underline{E}|$$

Hosoya and co-workers presented extensive tables with index Z for various acyclic[169] and cyclic graphs.[170] In the presentation of the index Z calculated in the following for I and II it will be observed that p(G,0) = 1, and p(G,1) = q, cf. Figure 2 on next page.

k	0	1	2
p(G,k)	1	7	11
Z	1 +	7 +	11 = 19

I

k	0	1	2	3
p(G,k)	1	6	10	4
Z	1 + 6 +		10 +	4 = 21

II

Analogously to the characteristic polynomial derived from matrix $\underline{A}$, Hosoya[171] introduced the distance polynomial

$$(-1)^n \det |\underline{D} - x \underline{\underline{E}}|$$

derived from the distance matrix $\underline{D}$.

Gutman et al.[172] introduced an index M_1 (which was also denoted occasionally[173] as Σ) based on the degrees v_i of vertices i in the hydrogen-depleted graph G. The degree of a vertex is the number of its adjacent vertices, or the number of edges incident with the given vertex.

$$M_1 = \sum_{i=1}^{n} v_i^2$$

$M_1 = 1^2 + 1^2 + 1^2 + 1^2 + 1^2 + 2^3 + 3^2 + 4^2 = 34,$ $M_1 = 1^2 + 1^2 + 2^1 + 2^2 + 2^2 + 2^2 + 2^2 = 22$

I   II

The sequence (which will be denoted by P) of vertex degrees for a given graph G may also be used for devising other topological indices, or for inducing an ordering of isomeric graphs, as will be indicated below under quadratic indices.

Two indices related to M_1 were devised on the numbers of vertices adjacent to all vertices of a path of length h. If h = 0, the path reduces to a vertex, and index M_1 is obtained. If h = 1, the path is an edge, and one obtains Randić's index[173] denoted by χ usually, but occasionally also denoted as M_2 by analogy with the preceding index.

$$\chi = M_2 = \sum_q (v_i v_j)^{-1/2}$$

Figure 2. Calculation of the Hosoya index Z for I and II

k	I p(G,k)	p(G,k) II
0	 1	1 
1	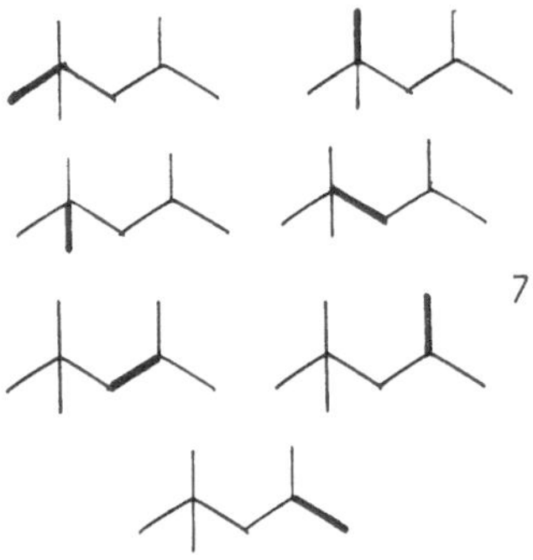 7	6 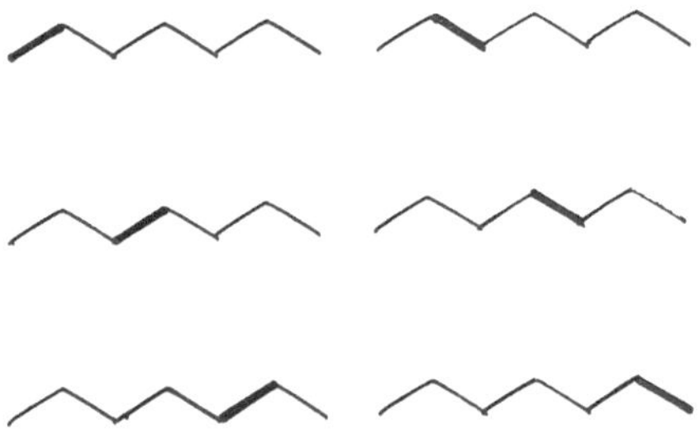
2	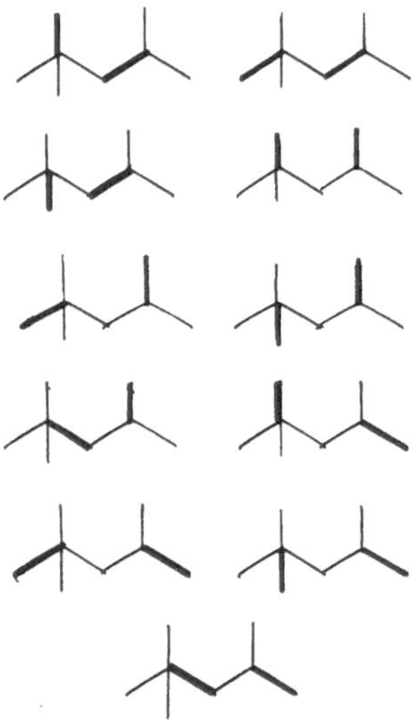 11	10 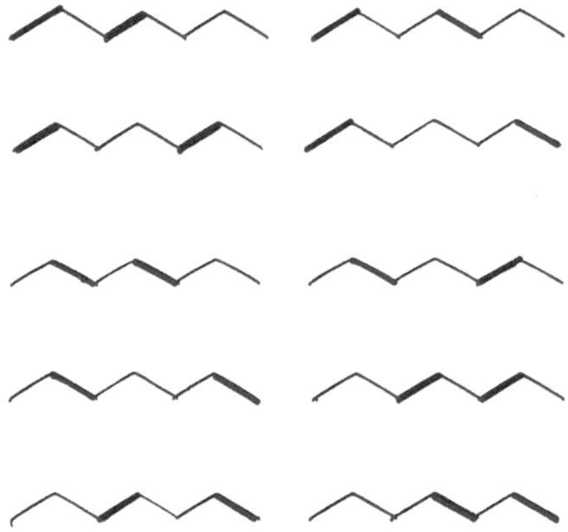
3		4 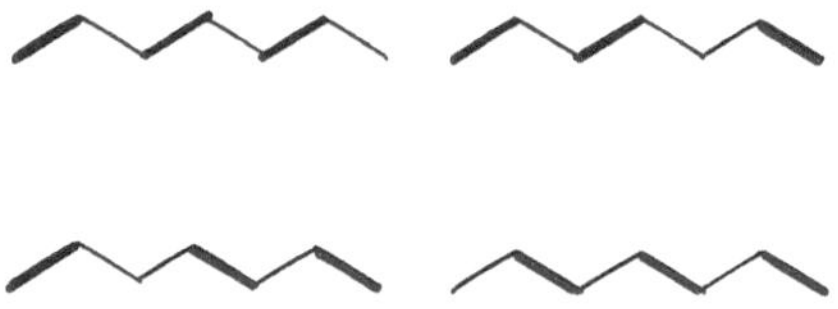
Z	1 + 7 + 11 = 19	1 + 6 + 10 + 4 = 21

v_i and v_j denote the degrees of the two endpoints of an edge in the hydrogen-depleted graph G, and the summation is extended over all q edges.

If the path has length h > 1 and includes vertices of degrees $v_1,\ldots,v_{h+1}$, Kier et al.'s generalized index[174,175] is obtained :

$$^h\chi = M_{h+1} = \Sigma\ (v_1 v_2 \cdots v_{h+1})^{-1/2}$$

An alternative definition for $^h\chi = M_{h+1}$ is from the h-th power of the adjacency matrix, i.e., from $\underline{A}^h$.[175] We present in the following the calculation of χ and $^2\chi$ for I and II. It should be noted that the paths of length 2 for the calculation of $^2\chi$ have been illustrated above as C-C-C fragments in discussing the Gordon index N_2.

Calculation of χ for I and II ($v_i v_j$ are given for each edge) :

$$\chi = 2\times3^{-1/2} + 3\times4^{-1/2} + 6^{-1/2} + 8^{-1/2} = 3.417;\quad \chi = 2\times2^{-1/2} + 4\times4^{-1/2} = 3.414$$

Calculation of $^2\chi$ for I and II :

$$^2\chi = 3\times4^{-\frac{1}{2}} + 2\times6^{-\frac{1}{2}} + 3\times8^{-\frac{1}{2}} + 3^{-\frac{1}{2}} + 24^{-\frac{1}{2}} = 4.159;\quad ^2\chi = 2\times4^{-\frac{1}{2}} + 3\times8^{-\frac{1}{2}} = 2.061$$

Lovasz and Pelikan[176] noted that the largest eigenvalue of the characteristic polynomial is a measure of branching, hence may be used as a topological index. For I, the largest eigenvalue is x_1 = 2.149, while for II, x_1 = 1.848.

Bonchev and Trinajstić[177] applied information theory to the problem of characterizing the molecular topology and succeeded in developing several new topological indices. According to Shannon's formula the information content I_o of a sequence of numbers X_i, where $\Sigma_i\ X_i = X$, is :

$$I_o = X\ \log_2 X - \Sigma_i\ X_i\ \log_2 X_i$$

and the mean information content is

$$\bar{I}_o = I_o/X = -\Sigma_i\ p_i\ \log_2 p_i$$

where $p_i = X_i/X$. For calculations it should be noted that

$$\log_2 a = (\log_{10} 2)^{-1} \log_{10} a = 3.32 \log_{10} a.$$

Thus, Bonchev, and Trinajstić defined the information content and mean information content on polynomial coefficients (I_{pc}, $\bar{I}_{pc}$), on realized distances in the graph G (I_D^W, $\bar{I}_D^W$), on the distribution of distances (I_D^E, $\bar{I}_D^E$). They also calculated the mean values of the indices of Hosoya ($\bar{Z}$), Wiener ($\bar{w}$), and Randić ($\bar{\chi}$). Formulas are :

$$I_{pc} = Z \log_2 Z - \sum_k p(G,k) \log_2 p(G,k)$$

$$\bar{I}_{pc} = \sum_k \frac{p(G,k)}{Z} \log_2 \frac{p(G,k)}{Z} = I_{pc}/k$$

$$I_D^E = \frac{n(n-1)}{2} \log_2 \frac{n(n-1)}{2} - \sum_i g_i \log_2 g_i$$

$$\bar{I}_D^E = - \sum_i \frac{2g_i}{n(n-1)} \log_2 \frac{2g_i}{n(n-1)} = I_D^E / \frac{n(n-1)}{2}$$

$$I_D^W = w \log_2 w - \sum_i g_i\, i\, \log_2 i$$

$$\bar{I}_D^W = - \sum_i g_i \frac{i}{w} \log_2 \frac{i}{w} = I_D^W/w$$

$$\bar{Z} = Z/K$$

$$\bar{w} = \frac{2w}{n(n-1)}$$

$$\bar{\chi} = \frac{\chi}{n(n-1)}$$

where K is the number of terms in the summation giving Z. The following page (Tab.8) lists the values of these indices for the two test systems I and II.

Because of the way they are defined, the above informational indices are more discriminating than the indices from which they are derived. For instance, Hosoya's indices for all heptane isomers have no degeneracies, but they have four degeneracies for octane isomers. However, the information content of the polynomial coefficients I_{pc} has no degeneracies for heptane or octane isomers.

Table 8. Informational indices for graphs I and II

Graph I Z=19, w=66, n(n-1)/2 = 28	Index	Graph II Z=21, w=56, n(n-1)/2 = 21
19 $\log_2 19$-7 $\log_2 7$-11 $\log_2 11$ = = 80.71-19.65-38.05= 23.01 bits	I_{pc}	21 $\log_2 21$-6 $\log_2 6$-10 $\log_2 10$ - - 4 $\log_2 4$= 92.24-15.51-33.22 - - 8.00 = 35.51 bits
$-\frac{7}{19}\log_2\frac{7}{19} - \frac{11}{19}\log_2\frac{11}{19}$ = = 7.669	$\bar{I}_{pc}$	$-\frac{6}{21}\log_2\frac{6}{21} - \frac{10}{21}\log_2\frac{10}{21}$ - $-\frac{4}{21}\log_2\frac{4}{21}$ = 8.877
28 $\log_2$ 28 - 7 $\log_2 7$ - - 10 $\log_2$ 10 - 5 $\log_2 5$ - - 6 $\log_2 6$ = 54.62 bits	I_D^E	21 $\log_2 21$-6 $\log_2 6$ - 5 $\log_2 5$ - -4 $\log_2 4$-3 $\log_2 3$ - 2 $\log_2 2$ = = 50.36 bits
$-\frac{7}{28}\log_2\frac{7}{28} - \frac{10}{28}\log_2\frac{10}{28}$ - $-\frac{5}{28}\log_2\frac{5}{28} - \frac{6}{28}\log_2\frac{6}{28}$ =1.951	$\bar{I}_D^E$	$-\frac{6}{21}\log_2\frac{6}{21} - \frac{5}{21}\log_2\frac{5}{21}$ - $-\frac{4}{21}\log_2\frac{4}{21} - \frac{3}{21}\log_2\frac{3}{21}$ - $-\frac{2}{21}\log_2\frac{2}{21}$ = 2.398
66 $\log_2 66$ - 40 $\log_2 2$ - - 30 $\log_2 3$ - 48 $\log_2 4$=307.16	I_D^W	56 $\log_2 56$ - 20 $\log_2 2$ - - 24 $\log_2 3$ - 24 $\log_2 4$ - - 20 $\log_2 5$-12 $\log_2 6$ = 233.46
$-\frac{7}{66}\log_2\frac{1}{66} - \frac{20}{66}\log_2\frac{2}{66}$ - $-\frac{15}{66}\log_2\frac{3}{66} - \frac{24}{66}\log_2\frac{4}{66}$ = = 4.654	$\bar{I}_D^W$	$-\frac{6}{56}\log_2\frac{1}{56} - \frac{10}{56}\log_2\frac{2}{56}$ - $-\frac{12}{56}\log_2\frac{3}{56} - \frac{12}{56}\log_2\frac{4}{56}$ - $-\frac{10}{56}\log_2\frac{5}{56} - \frac{6}{56}\log_2\frac{6}{56}$ =4.169
66/28 = 2.357	$\bar{w}$	56/21 = 2.667
19/3 = 6.333	$\bar{z}$	21/4 = 5.500
3.417/7 = 0.488	$\bar{\chi}$	3.414/6 = 0.569

Five new topological indices were proposed recently by Balaban[13] on the basis of sequences of numbers obtained on pruning (looping) a tree (tree-like) graph towards its center. A theorem by Jordan in graph theory states that any acyclic graph (tree) has a unique center (vertex) or bicenter (two adjacent vertices, i.e., two vertices joined by an edge). By pruning stepwise all vertices of degree one of a tree (end-points) together with their incident edges, one is left finally with the center or bicenter of the tree. The numbers δ_i of vertices deleted at each step constitute the "pruning sequence" of the tree, denoted by S. Evidently, $\Sigma_i \delta_i = n$; every one of the n vertices of the hydrogen-depleted graph G must be accounted for in the pruning sequence S. Another sequence of numbers encountered earlier was the sequence of vertex degrees; when arranged in nonincreasing order, the latter sequence is called the graphical partition of vertex degrees, and is denoted by P. It can be demonstrated that the sum of numbers in the sequence P is

$$\sum_{i=1}^{n} v_i = 2n - 2 .$$

Indeed, this sum is twice the number of edges, which in a tree is known to be n-1.

Analogously to index $M_1 = \sum_{i=1}^{n} v_i^2$ which, as indicated earlier, is based on the sequence P of vertex degrees (graphical partition of vertex degrees), the centric index $B = \sum_i \delta_i^2$ was proposed.[13] Its discriminating ability, though lower than that of Z, χ, or of informational indices, is higher than that of M_1 or of w. The higher is index B, the more branched is the alkane. A limitation of B originates in its definition (it can only be applied to acyclic systems); see, however, below.

In order to have topological indices which reflect the shape of an alkane, two "normalized" indices were defined[13] so as to be zero for the n-alkane (chain-graph): the normalized centric index C, and the normalized quadratic index Q. If the number of vertices of degree v_i is V_i, it can be shown[13] that indices M_1 and N_2 belong to the same class, which was termed the class of "quadratic indices", since $M_1 = \sum_i^n v_i^2$ and $N_2 = \frac{1}{2} \sum_i^n v_i(v_i-1)$. By adequate algebraic manipulation it can be evidenced that for graphs where all $v_i \leq 4$

$$M_1 = 16 V_4 + 9 V_3 + 4 V_2 + V_1 = 2(3 V_4 + V_3) + 4n + 6$$

$$N_2 = 6 V_4 + 3 V_3 + V_2 = 3 V_4 + V_3 + n - 2 .$$

The newly proposed[13] normalized indices are :

Normalized quadratic index

$$Q = 4 V_4 + \frac{3}{2} V_3 + 1 - \frac{1}{2} V_1 = 3 V_4 + V_3 = \frac{1}{2} \sum_i i V_i - 2n + 3$$

Normalized centric index

$$C = \frac{1}{2}(B - 2n + U) = \frac{1}{2} B(graph \; G) - B(chain\text{-}graph)$$

where $U = \left[1 - (-1)^n \right] / 2$.

The three quadratic indices Q, M_1 and N_2 are inter-related by the expressions :

$$Q = N_2 - n + 2 = M_1 / 2 - 2n + 3$$

$$M_1 = 2(N_2 + n - 1) .$$

In order to obtain "binormalized topological indices" with standard values both for the lower bound (namely, zero for the chain-graph) and for the upper bound (namely, one for the star-graph,which has the highest possible branching, and consists of one vertex of degree n - 1 bonded to n - 1 vertices of degree one), the following relations were proposed, by dividing the above indices with the values for the star-graphs :

Binormalized centric index $\qquad C' = \dfrac{B - 2n + U}{(n-2)^2 - 2 + U}$

Binormalized quadratic index $\qquad Q' = \dfrac{3 V_4 + V_3}{2(n - 2)(n - 3)}$

Values of all these topological indices for graphs I and II are presented in the adjoining Table 9.

As stated earlier in this review, for cyclic graphs no simple topological center could yet be found. However, quite recently, pro-posals for defining <u>polycenters</u> of any graph were presented by Bonchev, Balaban and Mekenyan [178]; from these proposals they derived centric topological indices for any graph (cyclic or acyclic) (as well as a

Figure 3. Finding the centers of graphs I and II, and the derived
pruning (lopping) sequence S

	Number of endpoints deleted on pruning step		
Graph I			Graph II

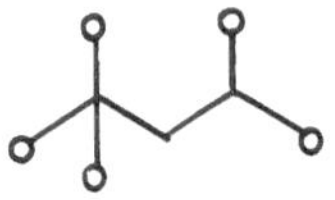	5	2	
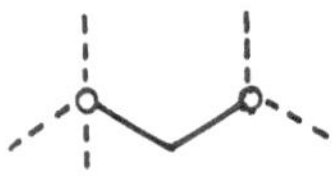	2	2	
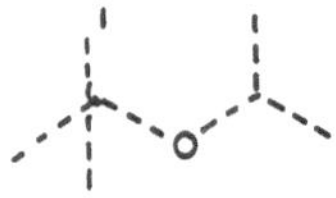	1	2	
		1	

Graph I		Graph II
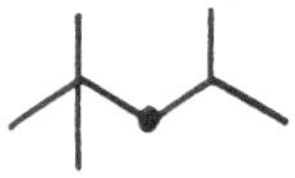	Centers of graphs	
5, 2, 1	Pruning sequence S	2, 2, 2, 1
4,3,2,1,1,1,1,1	Graphical partition P	2,2,2,2,2,1,1

Table 9. Centric and quadratic indices of graphs I and II

Graph I	Index	Graph II
$5^2 + 2^2 + 1^2 = 30$	B	$2^2 + 2^2 + 2^2 + 1^2 = 13$
$(30 - 16) / 2 = 7$	C	0
$3 \times 1 + 1 = 4$	Q	0
$7 / (6^2 - 2) = 0.412$	C'	0
$4 / (2 \times 6 \times 5) = 0.067$	Q'	0

new approach for topological codes which will not be discussed here).
These proposals are based on the distance matrix $\underline{D}$ of the graph G. For
obtaining the polycenter of a graph, its vertices are discriminated by
means of four criteria : (1) the smallest maximum distance in the row
or column of the vertex , (2) the smallest sum $\sum_i d_{ij}$ of the all distances d_{ij} in the row or column i corresponding to the given vertex ,
(3) the lowest number of largest entries in the 'distance partition'
of distances corresponding to the given vertex (this 'distance partition' is found by arranging the distances in the row or column of the
given vertex in nondecreasing order ; it is written in abbreviated
form as $1^s, 2^t, \ldots$ instead of $\underbrace{1,1,1,\ldots,1}_{s \text{ times}}, \underbrace{2,2\ldots,2}_{t \text{ times}}, \ldots)$, (4) the above

three criteria are applied in that order removing from the next criterion the vertices discarded by a previous criterion. After all three
criteria were applied, we are left with a pseudocenter formed by the
remaining vertices (and their incident edges, if these edges connect
vertices belonging to the pseudocenter). The first three criteria can
not discriminate further among the vertices of the pseudocenter. The
fourth criterion consists in submitting again to criteria (1) - (3)the
graph with the pseudocenter vertices, and in repeating this procedure
until on two successive cycles no more vertices can be removed, i.e.,
until on applying criteria (1) - (3) in this order to the pseudocenter
graph, the result is identical to the pseudocenter graph. This
'constant pseudocenter graph' is the polycenter of the graph : it may
contain one vertex or several adjacent or nonadjacent vertices. The
following page illustrates the procedure for finding the polycenter of
cyclic graphs : the first four examples have as polycenter a single
vertex, found by applying one, two, three, or all four of the above
criteria. In the figures, the polycenter is represented by black
points and the pseudocenter vertices which do not belong to the polycenter (they are removed by the fourth criterion) by white points. The
next examples illustrate how by minor structure variations the polycenter can change from a single vertex to a pair of disjoint or adjacent vertices. In some cases, the polycenter can have many vertices,
e.g., in a cyclic graph (cycloalkane) or in a complete graph (where
each vertex is linked to all other vertices) the polycenter includes
all vertices of the graph.

   Once the polycenter has been found, we can define the 'complete generalized partition' (CGP) of vertices according to their distances r_i from the polycenter vertex (vertices). Without entering de-

Table 10. Polycenters of cyclic graphs according to criteria (1)-(4)

Graph	Distance matrix 1 2 3 4 5 6 7 8	(1)	(2)	Criterion (3)	(4)
(triangle graph: 3, 2, 1, 4 with central black point)	1 ‖0 1 2 2‖ 2 ‖1 0 1 1‖ 3 ‖2 1 0 1‖ 4 ‖2 1 1 0‖	2 1 ← 2 2			
(graph III: 2,1,6,4,3,5) **III**	1 ‖0 1 2 2 3 2‖ 2 ‖1 0 1 1 2 1‖ 3 ‖2 1 0 1 2 2‖ 4 ‖2 1 1 0 1 2‖ 5 ‖3 2 2 1 0 1‖ 6 ‖2 1 2 2 1 0‖	3 2 2 2 3 2	6 ← 8 7 8		
(tree-cycle graph: 7,6,1,2,3,8,5,4)	1 ‖0 1 2 3 2 1 2 2‖ 2 ‖1 0 1 2 1 2 3 3‖ 3 ‖2 1 0 1 2 3 4 4‖ 4 ‖3 2 1 0 1 4 5 5‖ 5 ‖2 1 2 1 0 3 4 4‖ 6 ‖1 2 3 4 3 0 1 1‖ 7 ‖2 3 4 5 4 1 0 2‖ 8 ‖2 3 4 5 4 1 2 0‖	3 3 4 5 4 4 4 4	13 13	$1\,^2 2\,^4 3\,^2 1$ ← $1\,^3 2\,^2 3\,^2$	
(hexagon graph: 1,2,3,4,5,6 with center 4)	1 ‖0 1 2 1 2 1‖ 2 ‖1 0 1 2 2 2‖ 3 ‖2 1 0 1 1 2‖ 4 ‖1 2 1 0 1 2‖ 5 ‖2 2 1 1 0 1‖ 6 ‖1 2 2 2 1 0‖	2 2 2 2 2 2	7 8 7 7 7 8	$1\,^3 2\,^2$ $1\,^3 2\,^2$ $1\,^3 2\,^2$ $1\,^3 2\,^2$	5 5 3 ← 5
	1 ‖0 - 2 1 2 -‖ 3 ‖2 - 0 1 2 -‖ 4 ‖1 - 1 0 1 -‖ 5 ‖2 - 2 1 0 -‖				5 5 3 ← 5

Pseudocenters (white points) and polycenters (black points) of cyclic
graphs according to criteria (1) or (2) :

tails, we shall show how this is done taking graph III (Table 10) as an example. On the basis of the distance matrix and of the distance from the polycenter, we eliminate stepwise vertices according to (i) their distances from the polycenter and (ii) criteria (1) - (3) above. In graph III, vertex 5 is eliminated first since it is the only one at distance 2 from the polycenter vertex, i.e., r_5 = 2. Then vertex 1 is eliminated by criterion (1). Then vertices 3 and 6 follow according to criterion (2), then vertex 4, and finally the polycenter vertex 2; all four last vertices have $r_1 = r_3 = r_4 = r_6$ = 1 (for the polycenter, by de-finition, $r_2 = 0$). On the basis of the CGP = 1,1,2,1,1 which indicates the number of vertices pruned stepwise, in full analogy with the centric index for trees, we define the centric index $B_c = \sum_i^n r_i^2$ for cyclic graphs, i.e.,for III $B_c = 4 \times 1^2 + 1 \times 2^2 = 8$.

Also, the information content of the CGP leads to an inform-ational centric index $I_{c,r}$ for cyclic or acyclic graphs :

$$I_{c,r} = \Sigma r_i \log_2 \sum_i r_i - \sum_i r_i \log_2 r_i$$

Thus, for graph III, $I_{c,r}$ = 6 $\log_2 6$ - 4(1 $\log_2 1$) - 2 $\log_2 2$ = 13.51.

After this enumeration of the topological indices which have been proposed so far, we shall discuss some nonnumerical expressions which may also characterize the molecular topology, or may be used for inducing an ordering of graphs.

Some indices based on sequences of numbers may be expressed as polynomials, e.g., the Wiener index $w = \sum_i i g_i$ may be converted into Altenburg's expression depending on an indexed variable a_i, namely $\sum_i g_i a_i$, e.g., for graph I, $7a_1 + 10a_2 + 5a_3 + 6a_4$, and for graph II, $6a_1 + 5a_2 + 4a_3 + 4a_4 + 2a_5 + a_6$.

Smolenskii[179] expressed properties of alkanes in terms of the numbers and structural types of subgraphs : in a first-order approxi-mation, the property is expressed in terms of the numbers of three sub-graphs : path of length one, two, and three, denoted by X_1, X_2 and X_3, respectively. In a better approximation, the two latter numbers were decomposed into subclasses according to the degree of the vertices in the hydrogen-depleted graph G which are not endpoints of the path.This is exemplified for graphs I and II on the next page. Thus Smolenskii formalized in graph-theoretical language earlier work of Bernstein[180] and of Tatevskii, Papulov and co-workers.[181,182]

Figure 4. Derivation of Smolenskii's expressions for I and II

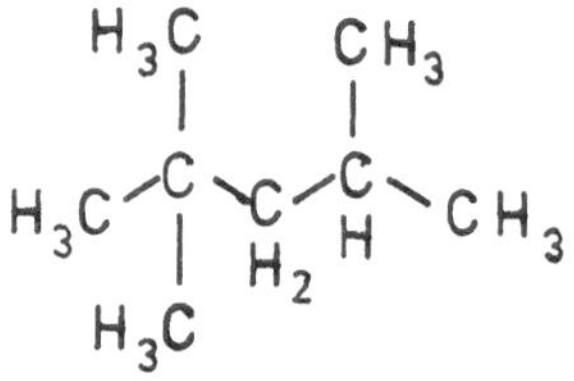

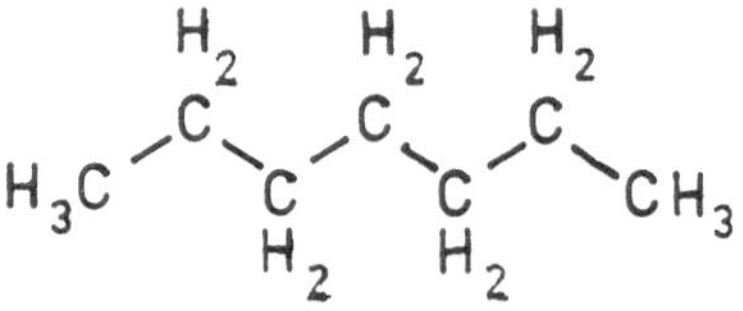

First approximation   No. of subgraphs

$X_1 = 7$

$X_2 = 10$

$X_3 = 5$

$X_1 = 6$

$X_2 = 5$

$X_3 = 4$

Second approximation

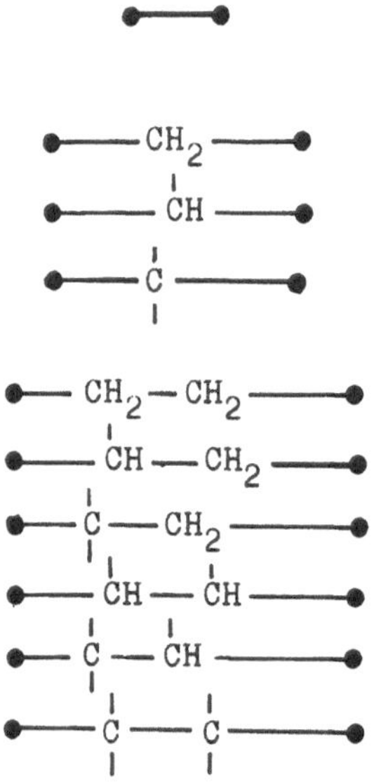

$X_1 = 7$

$X_1 = 6$

$X_2^1 = 1$ —CH$_2$—

$X_2^2 = 3$ —CH—

$X_2^3 = 6$ —C—

$X_2^1 = 5$

$X_2^2 = 0$

$X_2^3 = 0$

$X_3^1 = 0$ —CH$_2$—CH$_2$—

$X_3^2 = 2$ —CH—CH$_2$—

$X_3^3 = 3$ —C—CH$_2$—

$X_3^4 = 0$ —CH—CH—

$X_3^5 = 0$ —C—CH—

$X_3^6 = 0$ —C—C—

$X_3^1 = 4$

$X_3^2 = 0$

$X_3^3 = 0$

$X_3^5 = 0$

$X_3^5 = 0$

$X_3^6 = 0$

Evidently, Smolenskii's term X_2 is identical to Gordon and Scantlebury's index N_2 .

The sequence P of vertex degrees may be used for ordering isomeric graphs according to Muirhead's criterion for the comparability
of functions : [183,184] in two sequences of numbers in nonascending or-
der $a_1, a_2, \ldots, a_k$ and $a'_1, a'_2, \ldots, a'_k$ with $\sum_{i=1}^{k} a_i = \sum_{i=1}^{k} a'_i$, the former
sequence may be compared with the latter sequence and precedes it, if
for all i values in the range $1 \leq i \leq k$, Muirhead's inequality is
obeyed for every i = 1,2,...., k :

$$\sum_{i=1}^{i} a_i \geq \sum_{i=1}^{i} a'_i$$

If in at least one step during the comparison, the inequality becomes
reversed, the two graphs may not be compared, and are said to be in-
comparable.

The Figure 4 presents two comparable isomers of heptane (2-me-
thylhexane and heptane) : the former precedes the latter. Also, two
noncomparable 'isomers' of octane are presented, according to
Muirhead's criterion.

3.2 Applications of Topological Indices

In chemistry, topological indices were first used for correlating physical properties of chemical compounds. Platt[162] first at-
tempted topological correlations between the chemical structure of alkanes and such properties as molar refractions, molar volumes, boiling
points at normal pressure, or heats of formation. He showed that the
contributions of each C-C or C-H bond are influenced by 1st, 2nd, and
3rd C-C or C-H neighboring bonds. Soon afterwards, Wiener correlated
the same properties G by equations of type

$$G = \frac{a}{n^2} \sum_j j n_j + b n_3 = \frac{a}{w} + b \cdot p$$

where a and b are empirical adjustable parameters depending on the
property G, and n_j is the number of C-C bonds separated by j C-C bonds,
it is easy to prove that $n_1 = n - 1$; w and p are the Wiener numbers
mentioned earlier.

Later, [161] Wiener obtained good correlations with the same type
of equation for surface tension, specific dispersion, and critical solution temperature in aniline, of alkanes. For example we give below
the equation for the dependence of vapor pressure versus temperature

Figure 5. *Application of the Muirhead criterion for the comparability of functions to the ordering and comparability of graphs using the graphical partition P*

1. An example of comparable graphs

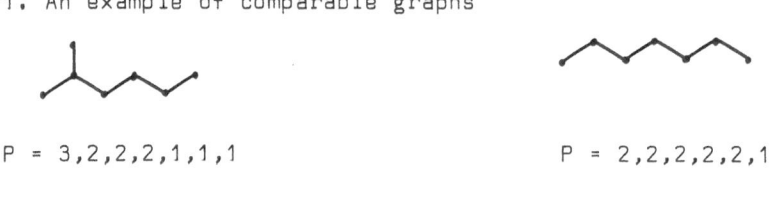

P = 3,2,2,2,1,1,1   P = 2,2,2,2,2,1,1

3	2	2	2	1	1	1
3	2	2	2	2	1	1

3 > 2

3 + 2 > 2 + 2

3 + 2 + 2 > 2 + 2 + 2

3 + 2 + 2 + 2 > 2 + 2 + 2 + 2

3 + 2 + 2 + 2 + 1 = 2 + 2 + 2 + 2 + 2

3+ 2 + 2 + 2 + 1 + 1 = 2 + 2 + 2 + 2 + 2 + 1

etc. = etc.

On the basis of Muirhead's criterion it results that 2-methylhexane is comparable with, and precedes, n-heptane.

2. An example of two noncomparable graphs

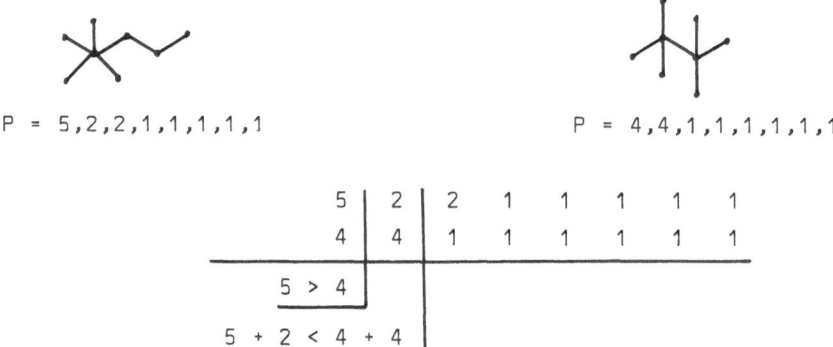

P = 5,2,2,1,1,1,1,1   P = 4,4,1,1,1,1,1,1

5	2	2	1	1	1	1	1
4	4	1	1	1	1	1	1

5 > 4

5 + 2 < 4 + 4

Since Muirhead's criterion is not satisfied, the two graphs with eight vertices are not comparable.

of alkanes : $\Delta t = \dfrac{98}{n^2} w + 5.5 \Delta p$, for isomers, n is constant.

In a critical analysis of such correlations, Platt[166] investigated three topological indices in multiparametric correlations :
p, w, and f. For heats of formation at absolute zero ($-273°C$) and for
molar refractions of alkanes, biparametric correlations with f and p
gave satisfactory results. For heats of vaporization, boiling points
at normal pressure, and molar volumes, biparametric correlations with
p and w proved adequate. For heats of formation of gaseous alkanes at
$25°C$, triparametric correlations had to be used.

The extensive data found in the literature (cf. Taylor et al.
[185] or Tatevskii et al.[181,182]) proved incentives for obtaining other
correlations. Smolenskii's expressions[179] for standard heats of formation of alkanes are (in a first approximation) :

$$\Delta H^O_{298} = 16.625 + 4.158\ X_1 + 1.074\ X_2 - 0.431\ X_3$$

A more accurate correlation is obtained with the expression :

$$\Delta H^O_{298} = 25.097\ X_2^1 + 10.775\ X_2^2 + 6.617\ X_2^3 - 20.158\ X_3^1 - 10.377\ X_3^2 -$$
$$- 7.038\ X_3^3 - 5.601\ X_3^4 - 3.950\ X_3^5 - 2.864\ X_3^6 .$$

Hosoya's index Z was extensively used for correlations.
Starting from the observation that both the size and the shape of alkanes influence the intermolecular forces (the larger, and the less
branched, alkanes boil at higher temperatures than the lower, and the
more branched, isomers or homologues), Hosoya proposed the following
relations for the boiling points (bp) of alkanes (in $°C$ at normal pressure) :

$$bp = (650.3\ \log Z + c)/\sqrt{n} + 46.55 .$$

Parameter c is adjustable for the type of substitution.
Hosoya et al. discussed the topological rules which apply to structure - boiling points relationships. However, the best correlation is
not a topological, but an empirical one :

$$\log(1078 - bp) = 3.0319 - 0.04999\ n^{2/3} .$$

Other correlations of Hosoya'a index Z in chemistry include correlations with calculated bond orders, and the use of Z as a sorting device for coding chemical structures.[165]

The centric indices, especially the binormalized ones, give good correlations with octane numbers for alkanes. However, biparametric correlations including also the number of carbon atoms, n, or the molecular weight of alkanes, give better results.[186]

Randić[187] as well as Burnham et al.[188] correlated diamagnetic susceptibilities of alkanes[187] and alkenes[187,188] with the mode of decomposition of the carbon skeleton into smaller subgraphs. Thus the general theory of diamagnetism of Hameka[189] is brought in correspondence with graph theory.

Randić's indices χ and $^2\chi$ were extensively used for chemical and biochemical correlations. Kier and Hall who developed such correlations reviewed many of these in a book which will be mentioned later.[190] We start with chemical correlations, and deal then at length with biochemical ones. Boiling points of alkanes with 2 - 7 carbon atoms show a good correlation with χ but the line is not straight, having a downward curvature. On the other hand, a very good straightline correlation is obtained between χ and the empirical branching index of Kováts, based on retention times in gas-liquid chromatography (vapor-phase chromatography).[191] Several other thermodynamical properties of alkanes correlate with χ : enthalpies of formation in the gas phase at $0^\circ C$ (ΔH_f^0), free energies, heats of solution, densities, refractive indices, parameters A, B, and C of the Antoine equation relating vapor pressures to temperatures in $^\circ C$. The illustrations on the next page exemplify the similarity between correlations with Z or with χ where n-alkanes give a straight line, and each isomeric series of alkanes forms a branching line of the main correlation.

Analogous to the correlations between boiling points and the Hosoya index Z, the correlation between ΔH_f^0 and χ gives within each group of isomeric alkanes straight parallel branches, emerging from the straight line correlating the data of n-alkanes. Thus,

$$H_f^0 = 12\chi - 11n - 9 \text{ (in kcal/mol)}$$

with a correlation factor r = 0.9998.

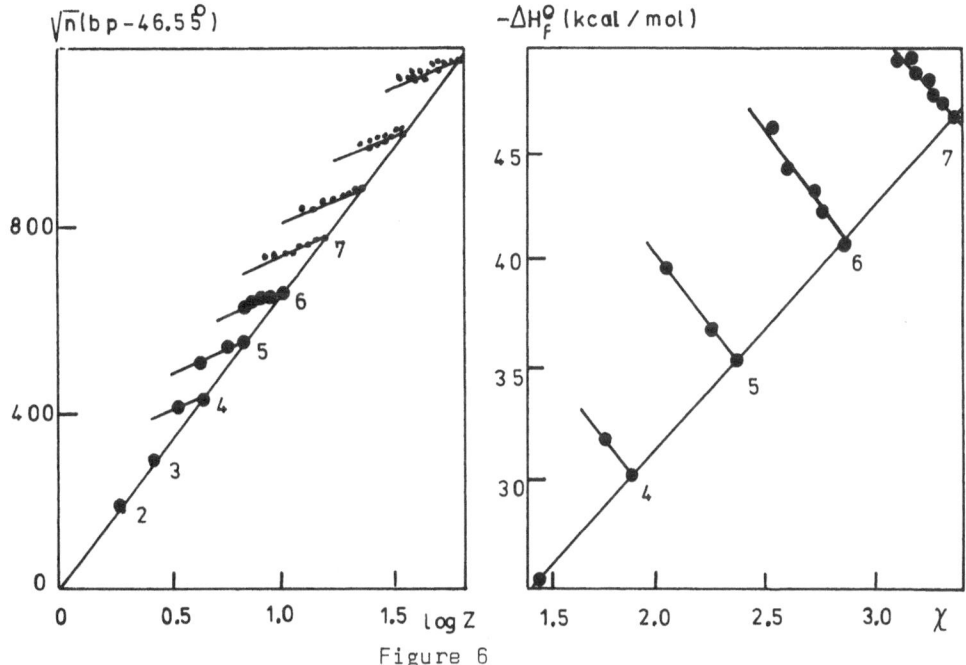

Figure 6

Finally, a good (though not linear) correlation is obtained between χ and the calculated molecular surface area, using Hermann's model. In turn, this molecular surface area gives good correlations with the solubilities of alkanes in water.

Nonlinear correlations between the informational index I_D^W, the number of carbon atoms n, and physical properties G of linear or branched polyalkylbenzenes gave excellent results. The correlations are of the form :

$$G = a + bn + cI_D^W + dnI_D^W$$

and the standard errors are : 0.04% for heats of formation, 0.07% for heats of combustion, 0.4% for molecular volumes or heats of vaporization; 1.3% for boiling points ; retention times for mono- or dialkyl-benzenes can be correlated with a standard error of 0.5%.[192]

The remainder of this chapter will be devoted to correlations between biological properties and topological indices. From all topological indices reviewed above, Randić's index χ was so far the index most tested in biological correlations, as indicated in Kier and Hall's book already mentioned.[190]

Kier[193,194] had shown that part of the drug-receptor interaction is due to long-range forces of dispersion and polarization. It is known that the polarizability of atoms in a molecule gives rise on one

hand to the above long-range forces, and on the other hand to bond polarizabilities. Indeed, bond and molecular polarizabilities were found
[193,195] to correlate with biological responses; now, since the atomic
polarizability of an atom relates to its volume, the total molecular
polarizability should be related to steric/topological properties of
the molecule.

In the first part of their series entitled "Molecular Connectivity", Kier, Hall, Murray, and Randić [196] found a significant
correlation between the calculated molecular polarizability of organic
compounds, α, and the index χ. The compounds were aliphatic alcohols,
ketones, esters, ethers, amines, phenols, heterocyclics, and hydrocarbons possessing anesthetic properties. In the following correlations, the regression equation will be followed by the correlation
coefficient r, the standard deviation s, and the number N of molecules
tested in the correlation :

$$\alpha = 1.60 + 9.26 \chi, \quad r = 0.990, \quad s = 3.59, \quad N = 36 .$$

Polarizabilities were calculated as sums of atomic refraction
constants, while in calculating χ for alcohols and hydrocarbons, the
hydrogen-depleted graph was drawn without differentiating carbon from
oxygen or other atoms with valency greater than one. For cyclic molecules which contain one extra bond for each ring relatively to acyclic
graphs, χ is calculated by subtracting the value of one bond : thus
cyclohexane and benzene are both considered to have $\chi = 2.500$. Two
properties were testes versus this index : the calculated cavity surface area (CSA) of aliphatic alcohols and hydrocarbons, and the minimum
blocking concentration of local anesthetics (MBC). Herman[147] studied
the solubility of organic solutes in water, and developed a procedure
for calculating the molecular cavity area (inside the solvent) which
accommodates the solute : this area contains the centers of the water
molecules in the first layer around the solute. Though this property
is essentially steric, it correlates well with Randić's topological
index :

$$CSA = 133.4 + 58.24\chi, \quad r = 0.978, \quad s = 11.2, \quad N = 69 .$$

Agin et al.[195] investigated, in a standardized set of conditions, the minimal concentration of a series of local anesthetics
for isolated muscle or nerve fibers. The MBC values found by Agin et al.

correlate well with χ in a semilogarithmic plot :

$$\log \text{MBC} = 3.55 - 0.762 \chi, \quad r = 0.983, \quad s = 0.390, \quad N = 36 .$$

In this correlation, acetanilide, aminopyrine, and chloroform were not included because their activities in the standard conditions or (for chloroform) the value of χ could not be determined.

Further correlations with biological activities were discussed in parts III, [198] IV, [199] and VI[200] of the series "Molecular Connectivity" by Kier, Murray, and Hall. Having found a linear correlation between χ and log P (where P is the partition coefficient of an organic compound between n-octanol and water, as determined by Hansch [2o1,202] and by Nys and Rekker [203]), it was natural to relate χ to biological properties of aliphatic alcohols, which had been found[204] to correlate linearly witg log P. Thus the following regressions were found, [198] (biological activities are expressed as A = p(Conc.)=- log (concentration for the biological effect) in all following regressions) :

Barnacle larvae narcosis : [205]

$$A = -1.167 (\pm 0.011) + 1.073(\pm 0.041)\chi, \quad r=0.991, \quad s=0.141, \quad N=15.$$

I_{100} movement of 2.5-day-old tadpoles at $18^\circ C$: [206]

$$A = -1.541(\pm 0.115) + 1.328(\pm 0.040)\chi, \quad r=0.997, \quad s=0.127, \quad N=8.$$

Idem for 12-day-old tadpoles : [198]

$$A = -1.41(\pm 0.088) + 1.294(\pm 0.030)\chi, \quad r=0.998, \quad s=0.097, \quad N=8.$$

For a series of aliphatic alcohols, ethers, ketones, esters, and carbamates, the following semilogarithmic correlation was found[199] between χ and the nonspecific narcotic effect on tadpoles (data [208]):

$$A = -0.931(\pm 0.158) + 0.922(\pm 0.049)\chi, \quad r=0.956, \quad s=0.297, \quad N=36.$$

Unsaturated compounds (ketones, esters) give better correlations if both bonds of the double bond are taken into account when calculating χ.

The regression between the nonspecific toxic effect on the Madison 517 fungus of monofunctional alcohols, ethers, ketones, and esters [199] and χ is :

$$A = 0.775(\pm0.032)\chi - 1.077(\pm0.119), \quad r=0.965, \quad s=0.263, \quad N=45.$$

A series of saturated and unsaturated aliphatic alcohols, ketones, phenols, aniline and pyridine inhibit the enzyme succinate oxidase from bovine muscle.[209] The correlation is : [199]

$$A = 0.916(\pm0.073)\chi - 1.582(\pm0.174), \quad r=0.966, \quad s=0.169, \quad N=13.$$

1-Alkyl- and -aralkyl-substituted thymidines inhibit the enzyme thymidine phosphorylase. The correlation between the inhibiting activity (the 50% inhibition concentration) and the index χ is [199] (data [210]):

$$A = 0.373(\pm0.051)\chi - 3.415(\pm0.325), \quad r=0.924, \quad s=0.207, \quad N=11.$$

Similarly, 1-alkyl-substituted adenines inhibit adenosine deaminase.[211] The correlation between the inhibiting activity and the index χ of the alkyl-substituted adenine is [199] (here, $A = \log I/S_{0.5}$) :

$$A = 0.449(\pm0.025)\chi - 2.081(\pm0.151), \quad r=0.991, \quad s=0.082, \quad N=8.$$

1-Decyl-3-piperidinecarboxamide-N,N-dialkyl derivatives inhibit butyrylcholinesterase. The correlation with index χ is :

$$A = 0.585(\pm0.025)\chi - 0.617(\pm0.241), \quad r=0.995, \quad s=0.062, \quad N=7.$$

Here A is pI_{50} for 50% inhibition.

The conclusion of these studies is that χ describes structural characteristics which govers drug-receptor interactions based only on geometric factors, excluding electrostatic interactions. The simplicity in calculating χ, and the fact that χ rests on purely nonempirical data, makes this index (or any other topological index) preferable to empirical parameters like the lipophilicity (log P), which necessitates experimental determinations and reference tables.

Not only linear, but also parabolic relationships were found between biological activities and log P values.[212,213] The fact that

log P and χ are linearly correlated makes it possible to obtain parabolic regressions

$$A = a\chi^2 + b\chi + c \; .$$

The Table 11 presents examples of such biological activities
which gave parabolic correlations with χ.[200] Only such correlations
are included which have at least six experimentally determined values,
and where the Fischer F test indicated that the addition of the χ^2
term to the linear regression in χ was significant at the 0.99 level
or better.

In all discussions above, topological indices were calculated
without paying attention to heteroatoms, assimilating them to carbon
atoms in the molecular chain. However, in their part III,[198] Murray,
Hall and Kier made the provision that each C-C bond must be counted in
the connectivity, e.g., in unsaturated compounds like $H_2C=CH-$, $HC\equiv C-$,
or $O=C\hspace{-0.3em}\diagdown$ compounds, the six C or O atoms indicated have connectivities
2, 3, 3, 4, 2, and 4, respectively; this requirement is not necessary
for polarizability correlations, but it is for correlations with solubility.

When the molecule does possess heteroatoms, for certain correlations their presence must be taken into account, in calculating
the topological index. In their part II,[214a] a term Q was introduced
into correlations with the convention that for hydrocarbons Q = 0,
and for alcohols Q = 1 (only these two classes were correlated for
water solubility). A second approach for alcohols was to assign the
C-OH bond different numerical values in primary, secondary, or ter-
tiary alcohols; as a result, the standard error for the logarithm of
the solubility decreased with 16%, and an analogous 33% decrease of
the standard error was obtained for the correlation of boiling points.
In their book,[190] Kier and Hall used a third method, namely to separate the branching due to the carbon skeleton from the branching
originating at the functional group; thus a 40% reduction of standard
errors was obtained for the heat of atomization of 20 saturated acyclic alcohols.

A more general method for including the presence of heteroatoms N or O in the calculation of the Randić index χ is to consider
their connectivity in the hydrogen-depleted graph as depending not
only on the number of bonds, but also on electrical charge and on the
nature of the heteroatom as indicated in Table 12. For halogens, however, empirically derived connectivity values (denoted by δ^V) have to

Table 11. Biological activities which present parabolic correlations with index X [199] and with log P. [212]

Compounds	Biological activity	Parameter[a]	Ref.
Alkyl-benzyl-dimethylammonium salts	Inhibition of P. aeruginosa	MIC	204
"	" " S. aureus	"	"
"	" " Cl. welchii	"	"
"	Toxicity towards P. aeruginosa	MKC	"
"	" " Cl. welchii	"	"
"	Hemolysis of sheep red cell	CH_{50}	"
5,5-Dialkylbarbituric acids	Narcotic effect in rabbit	MHD	"
"	" " " mouse	AD_{50}	"
"	" " " rat	MHD	"
"	" " " rat brain	I_{50}	"
Alkyl or aralkyl carbamates	In vitro intestinal absorption(serosal transfer)	Rate	213
"	" " " (tissue bound)	"	"
"	" " " (mucosal loss)	"	"
Aryl-substituted alkyl-benzyl-di-methylammonium salts	Toxicity towards S. aureus	MKC	204
Alkoxy-anilines, -aminopyridines, -aminopyrimidines and α-naphthylamines	Inhibition of M. tuberculosis	MIC	"

[a] Abbreviations : MIC = minimum inhibitory concentration; MKC = minimum killing concentration ;
CH_{50} = minimum concentration for 50% hemolysis; MHD = minimum hypnotic dose ;
AD_{50} = minimum anesthetic dose for 50% population; I_{50} = concentration for 50% inhibition.

be employed in the calculation of index χ^V, the sum extending over all edges

$$\chi^V = (\delta_1^V \, \delta_2^V)^{-1/2}$$

Thus, δ^V-values for halogens are : F, -20; Cl, 0.69; Br, 0.254;I,0.085.

Table 12. Values of connectivities δ^V for C, N, and O atoms [214b]

Atom X :	Carbon	Nitrogen		Oxygen	
No. of C-X bonds Charge:	0	+1	0	+1	0
0	CH_4 0	NH_4^+ 1	NH_3 2	OH_3^+ 3	OH_2 4
1	$-CH_3$ 1	$-NH_3^+$ 2	$-NH_2$ 3	$-OH_2^+$ 4	$-OH$ 5
2	$=CH_2$ 2	$=NH_2^+$ 3	$=NH$ 4	$=OH^+$ 5	$=O$ 6
3	$\equiv CH$ 3	$\equiv NH^+$ 4	$\equiv N$ 5		$\equiv O^+$ 7
4	$\equiv \underline{C}$ 4		$\equiv \underline{N}^+$ 6		

It should be mentioned that table 12 includes also a few values which were not present in the original paper, and that it rationalizes the values in terms of electrical charge, nature of atom, and number of simple or multiple C-X bonds.

The resulting correlations are appreciably improved relatively to the results when the heteroatoms has been assimilated to carbon atoms. Thus for boiling points (bp) at normal pressure, one obtains instead of the initial regression which assimilated all heteroatoms with carbons :

$$bp = 38.79(\pm 1.61)\chi + 11.26(\pm 5.85), \quad r=0.963, \quad s=8.39^o, \quad N=48$$

a new, improved regression :

$$bp = 196.58(\pm 11.34)\chi - 157.6(\pm 11.31)\chi^V - 41.24(\pm 4.56),$$

$$r=0.993, \quad s=3.68^o, \quad N=48.$$

For the logarithm of the molar solubility of organic compounds in water, S, instead of the initial regression :

$$\log S = -2.61(\pm 0.086)\chi + 6.52(\pm 0.31), \quad r=0.976, \quad s=0.447, \quad N=48$$

one obtains a new, improved correlation :

$$\log S = 9.27(\pm 0.98)\chi + 6.64(\pm 0.97)\chi^V + 8.73(\pm 0.39),$$

$$r=0.998, \quad s=0.317, \quad N=48,$$

with the additional benefit of destroying redudancies and ordering consistently the boiling points and solubilities in relation to structure.

For the diols and triols, instead of the initial regression :

$$bp = 20.71(\pm 24.28)\chi + 159.38(\pm 67.23), \quad r=0.307, \quad s=34°C, \quad N=9,$$

one obtains a tenfold improvement in the standard deviation s :

$$bp = 294.64(\pm 10.04)\chi - 222.1(\pm 9.38)\chi^V - 29.11(\pm 10.92),$$

$$r=0.995, \quad s=3.78°, \quad N=9,$$

Primary amines, secondary amines, and alkyl halides give separate correlations for boiling points.

Primary amines : instead of

$$bp = 50.93(\pm 1.00)\chi - 49.17(\pm 3.30), \quad r=0.996, \quad s=4.59°, \quad N=21,$$

one obtains

$$bp = 154.09(\pm 17.20)\chi - 103.5(\pm 17.84)\chi^V - 75.92(\pm 5.03),$$

$$r=0.999, \quad s=2.79°, \quad N=21 .$$

Secondary amines : instead of

$$bp = 48.50(\pm 1.15)\chi - 58.55(\pm 4.07), \quad r=0.997, \quad s=4.28°, \quad N=13,$$

one obtains

$$bp = 171.40(\pm 21.84)\chi - 120.07(\pm 21.45)\chi^V - 102.4(\pm 8.07),$$

$$r=0.999, \quad s=2.20°, \quad N=13.$$

For alkyl halides, one obtains with

$$bp = 19.91(\pm 2.41)\chi + 38.07(\pm 1.27)\chi^V - 69.55(\pm 5.17),$$

$$r=0.992, \quad s=4.79^0, \quad N=24.$$

Another property which can be successfully correlated with χ is the molar refraction. For a group of substituents including alkyls, ethers, amines, esters, amides, ketones (and one aldehyde, alcohol, and nitroderivative), the increments R_m of molar refractions are calculated to be

$$R_m = 2.656(\pm 0.593)\chi + 7.14(\pm 0.688)\chi^V - 0.958(\pm 0.518),$$

$$r=0.990, \quad s=1.03, \quad N=25,$$

and if haloalkyls are also included in the above correlation,

$$R_m = 3.661(\pm 0.227)\chi + 5.23(\pm 0.211)\chi^V + 4.460$$

$$r=0.990, \quad s=1.022, \quad N=65.$$

4. MSD, MINIMAL STERIC DIFFERENCE (SIMPLE VERSION)

It is obvious that steric fit depends on the shape of both the biological receptor and the effector molecule. Steric parameters described in the previous chapters rely only on the shape of the effector molecule. The regressional equation will indicate if the receptor prefers large or small values for the corresponding parameter (for example van der Waals volume of the substituent or molecule, width, length, branchedness etc.); at most the optimal value of the shape characteristic corresponding to the steric parameter is indicated.

Minimal steric difference[14-16], MSD, depends both on the shape of the effector molecule and of the biological receptor. The basic idea of the minimal steric difference concept is that the affinity of effectors for a receptor is a linearly decreasing function of the sum of nonoverlappable volumes of effector molecule and receptor cavity. In order to have this concept at work, one must have a guess for the shape of the receptor cavity and a simple method to evaluate the nonoverlappable volumes - which is the MSD. Thus, if A_i is the biological activity of molecule M_i, for correlational equations it should depend on MSD_i as :

$$\hat{A}_i = \alpha - \beta \ MSD_i \tag{1}$$

A first guess for the receptor cavity shape will be the natural effector or the molecule of highest activity known. This molecule with the shape (approximately) complementary to the receptor cavity is called "standard"(S). One seeks the maximal superposition of the molecule M_i upon S. The number of nonsuperposable (nonhydrogen) atoms from both M_i and S gives MSD_i. Hydrogen atoms are neglected in order to simplify the problem, since their van der Waals volumes and covalent radii are rather low. Thus atom X and groups - XH, $-XH_2$ etc. are considered equivalent. As heavy atoms have larger van der Waals volumes one can attribute a weight 1 to First Row (second period) atoms (C,N,O), 1.5 to Second Row (third period) atoms (S,Cl) and 2 to higher period atoms (Br,I). If a higher period atom (for example I) is superimposed on a lower period atom (for example CH_3) one will add the difference of weights to MSD, but if a higher period atom is superimposed on a trifurcation (for example I on CH of $-CH(CH_3)_2$) one will subtract the weight of the higher period atom from the sum of the weights of the trifurcation atoms. Thus the I on CH_3 superposition will add 2-1=1 to MSD, the I on $-CH(CH_3)_2$ superposition 3-2=1, leading in both cases to the same increment.

In the superposition process, differences between bond length and bond angles are neglected. One must avoid to superimpose pairs of atoms bonded in one molecule on nonbonded pairs of atoms in the second one, as van der Waals contact distances are much larger (3-4 Å) than covalent bonds (1.2-2 Å). If the molecule M_i has several low energy conformations it will enter the receptor with the conformation which fits best the cavity : one must consider for M_i as MSD_i the conformation which gives maximal superposition on S, i.e., one must select the minimal steric difference. Low energy conformations will generally be considered all those without obvious distorsions of bond angles or van der Waals repulsions of nonbonded atoms (groups). Sometimes certain atoms within M_i will have to be superimposed on a certain group of atoms from S : for example in enzymatic reactions, the atoms of bonds to be cleaved in M_i and in S must be superimposed one upon another, respecting also the corresponding orientation of bonds.Groups suspected to give strong interactions with the receptor, for example electrically charged groups, should also be suitably superimposed within the M_i on S.

As examples, some superpositions on $R-CH_2-C_6H_5$ as standard,S, are depicted in Figure 7. Both S and M are always depicted, unsuper-

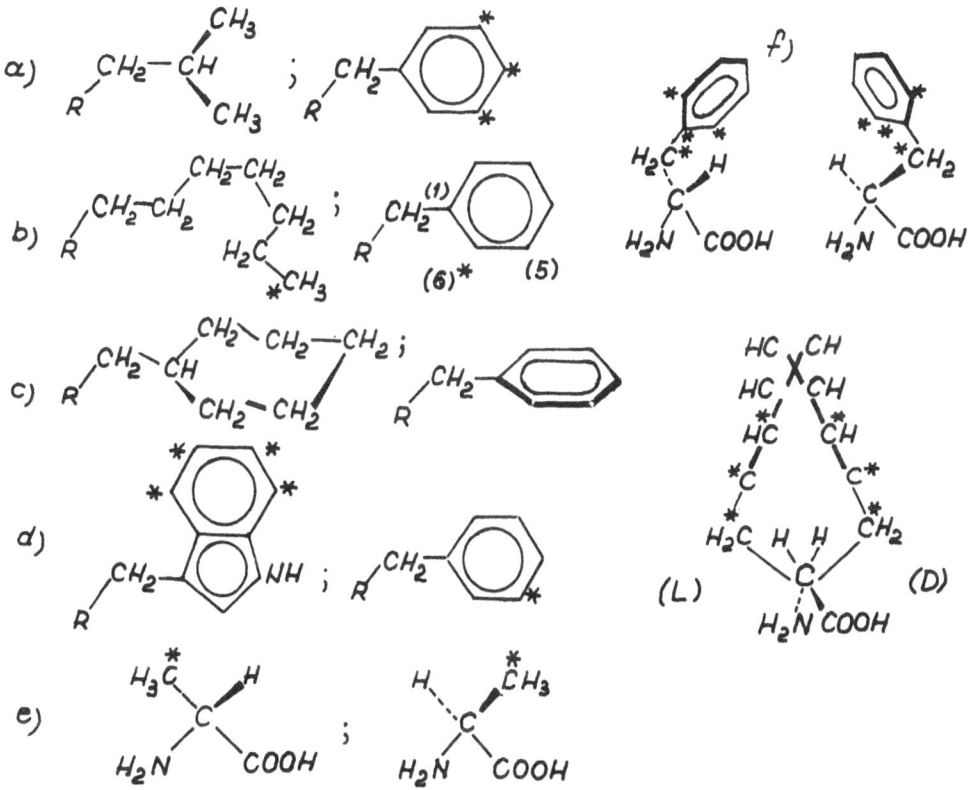

Figure 7. Examples of superpositions and calculation of MSD
M_a=isobutyl-R, MSD_a=3; M_b=n-heptyl-R, MSD_b=4; M_c=methylene-cyclohexyl-R, MSD_c=0; M_d=methylene-β-indolyl R, MSD_d=5; for M_a, M_b, M_c, M_d, S is $RCH_2C_6H_3$. For cases e) and f) D-aminoacids are superimposed on L-amino-acids; for alanine, MSD_e=2, for phenylalanine MSD_f=8.

imposable atoms are marked by asterisks. The isobutyl, n-hexyl and cyclohexyl methylene groups are considered in approximately planar conformations; the corresponding MSD's are MSD_a=3, MSD_b=2, MSD_c=0, MSD_d=5, MSD_e=1, MSD_f=8. The end CH_2-CH_3-group of n-heptyl cannot be superimposed on positions 5,6 of the phenyl cycle as the nonbonded and CH_3 and 2-nd CH_2 group of n-heptyl would be superimposed on the bonded C atoms 6 and 1 of the phenyl cycle. The superposition (d) of the five-membered ring moiety of indolyl on phenyl (six-membering ring) con-tributes 1 to MSD. In superposing D- on L-phenylalanine, the phenylic cycles cannot be completely superposed : this would require C_α, CH_2, C_1 and C_6 coplanar, with steric repulsion between H_α and H_6, analogous to biphenyl which is not coplanar.

4.1 *Correlations biological activity - MSD*

The simple version, MSD is especially suited for correlations
involving aminoacid replacements in oligopeptides, where the natural
effector can be taken as standard. For this purpose, Table 13 lists
MSD's obtained by substitutions of the 20 aminoacids among them, both
L and D isomers. Superpositions of aminoacids are illustrated in
Figure 8.

Decrease of oxytocitic activity was correlated with MSD in
the first paper of the series[14]. The decrease, ΔA_i, is calculated with
respect to the activity of oxytocine :

$$\overline{CysS-Tyr-Ile-G1N-AsN-CysS}-Pro-Leu-Gly \quad (9)$$

which was used also as standard for MSD calculations.

The MSD values are listed in Table 14. Although only r=0.545
is obtained, the correlation is highly significant for N=42 compounds.
Steric fit is not the only factor in determining effector-receptor affinity and even the receptor-rigidity will not be the same for all
aminoacidic positions in the peptide.

Another activity - MSD correlation[215] was presented on a series of N=18 decapeptides with luteinizing hormone releasing activity,
as studied by Fujino et al.[215]. The values for correlation are listed
in Table 15. The best molecule (R-NHCH$_2$CH$_3$, with R: Glu-His-Trp-Ser-Tyr-
Gly-Leu-Arg-Pro) is considered as standard and the following correlational equation is obtained :

$$\hat{A}_i = 2.245-0.535 \, MSD_i, \quad r=0.772, \quad s=0.596, \quad N=18 \qquad (2)$$

The toxicites of nine tionphosphoric acid esters (Table 16)
were correlated with MSD, using the best compound (Nr.1 of Table 16)
as standard[217].
Enzymatic oxydation of the P=S group to a P=O group seems to determine the toxicity of these compounds[219]. Activities A in Table 16
are -log LD$_{50}$ (lethal dosis, 50%) for rats. This correlation yields
r = -0.85 and β=0.36.

Results of other A-MSD correlations are summarized in Table 17.
The best compounds of the series was used as standard for correlations nr.2, 3 and 4, the natural effector (acetylcholine) for 6 and 7,

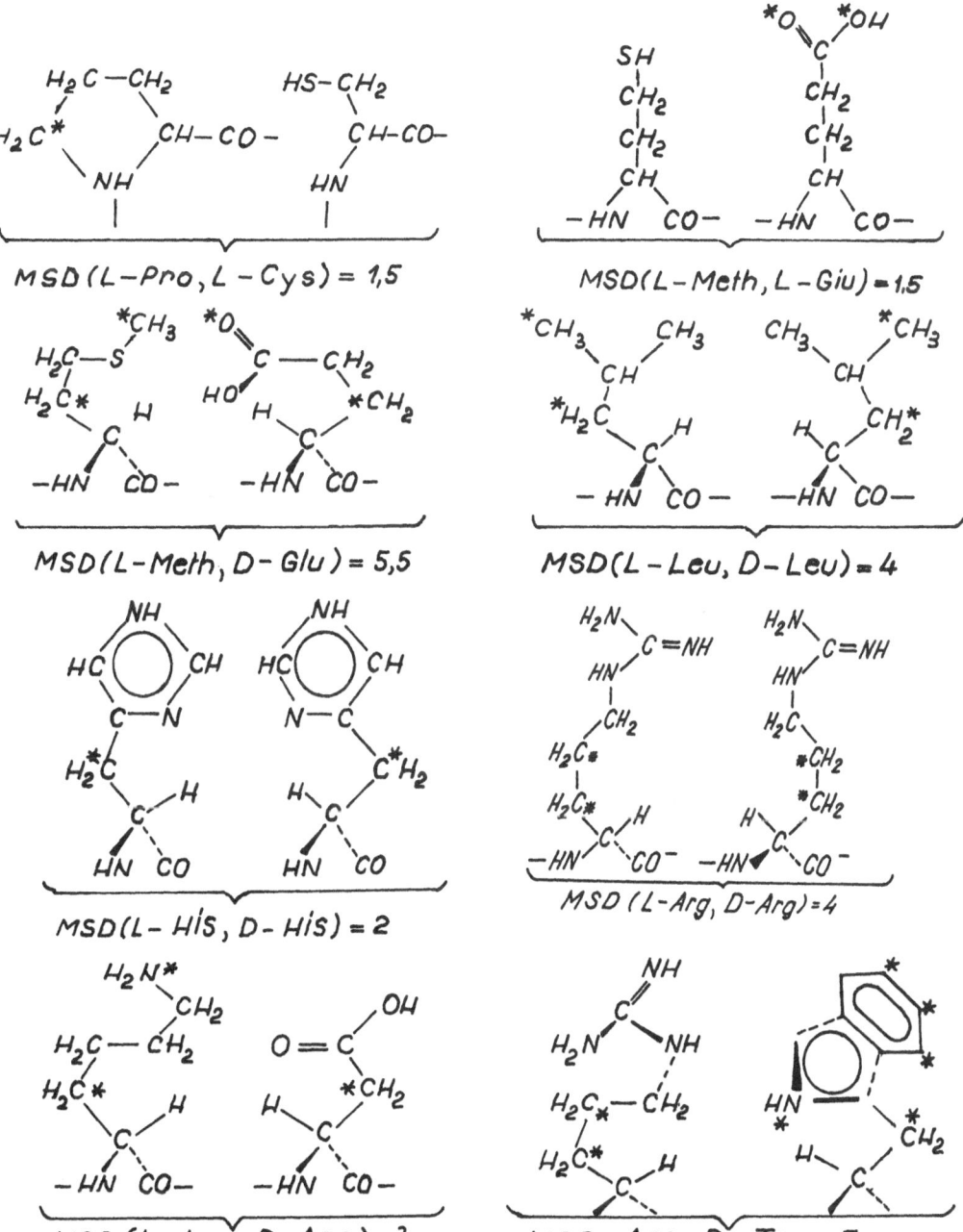

Figure 8. Calculation of MSD's for some replacements of aminoacids in a peptide

Atoms (XH_n groups) with asterisk are unsuperimposable in the pairs for which the corresponding MSD is calculated

Table 13. MSD's for aminoacid - aminoacid replacements in peptides

	L-Arg	L-Lys	L-Glu	L-Asp	L-GlN	L-AsN	L-Thr	L-Ser	L-His	L-Trp	L-Tyr
Gly	7	5	5	4	5	4	3	2	6	10	8
L-Ala	6	4	4	3	4	3	2	1	5	9	7
L-Val	6	4	4	3	4	3	0	1	5	9	7
L-Leu	5	3	3	0	3	0	3	2	2	6	4
L-Ile	5	3	3	2	3	2	1	2	4	8	6
L-Cys	4.5	2.5	2.5	1.5	2.5	1.5	1.5	0.5	3.5	7.5	5.5
L-Met	3.5	1.5	2.5	2.5	2.5	2.5	3.5	2.5	2.5	6.5	4.5
L-Pro	6	4	4	3	4	3	2	1	3	7	5
L-Phe	2	2	4	3	4	3	6	5	1	4	1
L-Tyr	1	3	5	4	5	4	7	6	2	3	[10]
L-Trp	5	5	7	6	7	6	9	8	4	[12]	14
L-His	3	1	3	2	3	2	5	4	[2]	6	10
L-Ser	5	3	3	2	3	2	1	[4]	6	12	10
L-Thr	6	4	4	3	4	3	[6]	5	7	11	11
L-AsN	5	3	3	0	3	[4]	5	4	4	8	10
L-GlN	4	2	0	3	[6]	5	6	5	6	13	10
L-Asp	5	3	3	[4]	5	4	5	4	4	8	10
L-Glu	4	2	[6]	5	6	5	6	5	6	13	10
L-Lys	2	[4]	5	3	5	3	5	5	3	9	7
L-Arg	[4]	6	7	5	7	5	7	7	5	7	7

Table 13 (continued)

L-Phe	L-Pro	L-Met	L-Cys	L-Ile	L-Leu	L-Val	L-Ala	Gly	
7	3	4.5	2.5	4	4	3	1	[0]	Gly
6	2	3.5	1.5	3	3	3	[2]	1	D-Ala
6	2	3.5	1.5	1	3	[6]	4	3	D-Val
3	3	2.5	1.5	2	[4]	5	5	4	D-Leu
5	3	2.5	1.5	[4]	4	5	5	4	D-Ile
4.5	1.5	3	[5]	4.5	4.5	5.5	3.5	2.5	D-Cys
3.5	3.5	[3]	4	3.5	2.5	4.5	5.5	4.5	D-Met
4	[6]	7.5	5.5	7	7	6	4	3	D-Pro
[8]	10	7.5	9.5	9	9	10	8	7	D-Phe
9	11	8.5	10.5	10	10	11	9	8	D-Tyr
13	13	7.5	10.5	10	8	11	11	10	D-Trp
9	9	3.5	6.5	6	4	7	7	6	D-His
9	5	4.5	4.5	4	4	5	3	2	D-Ser
10	6	5.5	5.5	5	5	6	4	3	D-Thr
9	7	2.5	5.5	4	4	5	5	4	D-AsN
9	8	5.5	4.5	5	5	6	6	5	D-GlN
9	7	2.5	5.5	4	4	5	5	4	D-Asp
9	8	5.5	4.5	5	5	6	6	5	D-Glu
6	6	5.5	5.5	5	3	5	6	5	D-Lys
6	7	7.5	7.5	7	5	7	8	7	D-Arg

MSD's between the L-aminoacids written on top and the left
column L-Aminoacids are given above the diagonal; between
the same L and D aminoacids, on the diagonal; between L-aminoacids
and the right column D aminoacids, below the diagonal.
The C_α, -NH-, -CO-groups occupy always the same positions in
both the aminoacids of the pair.

Table 14. *Activity decrease (ΔA) correlations in oxytocine amino-acid replacement derivatives*

Nr.	Replacement	MSD	ΔA	Nr.	Replacement	MSD	ΔA
1	Phe_2	1	1.15	23	Ser_2His_3	7	5.0
2	Ser_2	6	5.0	24	Ser_2Lys_8	9	5.0
3	Phe_3	5	1.25	25	His_2Phe_3	7	5.0
4	Tyr_3	6	3.65	26	Phe_3Lys_8	8	1.95
5	Trp_3	8	4.05	27	Phe_3Arg_8	10	0.8
6	Leu_3	2	1.0	28	Phe_2His_8	7	2.4
7	Val_3	1	0.9	29	Tyr_3Lys_8	9	4.6
8	Ser_4	3	0.3	30	Trp_3Lys_8	11	5.0
9	Ala_4	1	1.1	31	Ser_4Ile_8	5	0.5
10	AsN_4	3	0.6	32	AsN_4GlN_5	7	3.05
11	Ser_5	2	2.8	33	$Phe_2Phe_3Lys_8$	9	3.2
12	Ala_5	3	5.0	34	$Phe_2Phe_3Arg_8$	11	3.35
13	GlN_5	3	2.65	35	$Phe_2Tyr_3Lys_8$	10	5.0
14	Val_5	3	5.0	36	$Ser_2His_3Lys_8$	10	5.0
15	Ile_8	2	0.2	37	$His_2Ser_3Lys_8$	7	5.0
16	Val_8	3	0.35	38	$His_2Phe_3Lys_8$	10	5.0
17	Lys_8	3	0.75	39	$Phe_3AsN_4Lys_8$	11	2.2
18	Arg_8	5	0.75	40	$Phe_3Ala_4Lys_8$	9	2.95
19	Phe_2Phe_3	6	2.1	41	$Phe_3Ser_4Lys_8$	11	2.7
20	Phe_2Tyr_3	7	5.0	42	$Phe_3Ser_5Lys_8$	10	5.0
21	Phe_2Lys_8	4	2.65				
22	Phe_2Arg_8	6	2.35		− r = 0.545		

Table 15. *LH releasing activity of LHRH analogues with modified tenth aminoacidic residue, R*

Nr	R	MSD	A	Nr	R	MSD	A
1	$-N-CH_2-C-NH_2$ (H...O)	2	2.00	10	$-N(CH_2)_2$	2	1.17
2	$-NH_2$	2	1.04	11	cyclo $N(CH_2)_4$	2	2.04
3	$-NHCH_3$	0	2.48	12	cyclo $N(CH_2)_4O$	3	1.23
4	$-NHCH_2CH_3$	0	2.48	13	cyclo $N(CH_2)_5$	3	0.18
5	$-NHCH_2CH_2CH_3$	1	2.28	14	$-Pro$	5	-0.75
6	$-NHCH_2CH_2OH$	1	2.32	15	$-AlaNH_2$	3	0.78
7	$-NHCH_2CH_2CH_2CH_3$	2	0.86	16	$-NHCH(CH_3)_2$	1	2.18
8	$-OCH_3$	1	0.79	17	$-NHCH_2CH(CH_3)_2$	2	1.48
9	$-OCH_2CH_3$	0	1.06	18	$-NHCH(CH_2)_5$	4	-0.15

Table 16 *Toxicity of thionphosphoric acid esters*

Nr.	Compound	A	MSD
1.	$(C_2H_5O)_2P(S)-O$—⬡—NO_2	- 0.85	0
2.	$(CH_3O)_2P(S)-O$—⬡—NO_2	- 1.00	2
3.	$(C_2H_5O)_2P(S)-O$—⬡—NO_2, CH_3	- 1.00	1
4.	$(CH_3O)_2P(S)-O$—⬡—NO_2, CH_3	- 2.40	3
5.	$(CH_3O)_2P(S)-O$—⬡—NO_2, Cl	- 2.40	3.5
6.	$(CH_3O)_2P(S)-O$—⬡—SCH_3	- 1.00	2.5
7.	$(CH_3O)_2P(S)-O$—⬡—SCH_3, Cl	- 2.40	3.5
8.	$(C_2H_5O)_2P(S)-O$—⬡—Cl, Cl	- 2.13	4.5
9.	$(CH_3O)_2P(S)-O$—⬡—Cl, Cl	- 3.00	6.5

Table 17 *Different correlations with MSD*

Nr.	Biological activity (effector molecules)	N	r
1.	α-Chymotrypsine catalyzed hydrolysis (methyl esters of N-acetylated L-aminoacids)[14]	14	0.951
2.	Dihydrofolate reductase inhibition (di-2,4-NH_2-pyrimidines, 5-subst.,aliphatic, cycloaliphatic[16]	11	0.937
3.	Glycoside C-N bond hydrolyses (purine and pyrimidine ribosides)[16]	10	0.571
4.	Adenosinekinase transphosphorylation (purine and pyrimidine ribosides)[16]	11	0.626
5.	α-Chymotrypsine catalysed hydrolyses (p-Nitrophenolic esters of carboxylic acids and D and L aminoacids)[219]	42	0.448
6.	Toxicity for mammals (organophosphorus compounds of the Schrader type)[220]	51	0.327
7.	Inhibition of acetylcholinesterase (organophosphorus compounds of the Schrader type)[221]	60	0.717
8.	Benzene: water partition (various organic compounds)[140]	63	0.037
9.	Ethyl ether: water partition (various organic compounds)[222]	41	0.039
10.	Affinity for various antipurinoyl bovine serum albumin sera (purine and pyrimidine derivatives)[223]	14	0.81 - 0.97

a standard improved by trial and error for 1,5 and 10 and the non-aqueous solvent for 8 and 9. In the case of correlations 8 and 9 it seems that solvent : water partition does not depend on steric similarity with the nonaqueous solvent, which may suggest that passive transport across biological membranes does not depend on any sort of steric fit.[222]

4.2 Correlations with MSD and Other Parameters

It is obvious that effector-receptor affinity depends not only on steric fit but also on solvation (hydrophobic effects), juxtaposition of groups among which strong intermolecular forces appear and, if the biological activity implies an effector-receptor reaction, also on σ-Hammett or other electronic reactivity parameters. Several multiparametric correlations including MSD were performed by our group, and an account thereof is given in the following paragraphs.

4.2.1 Oxytocic Activity of Oxytocine Aminoacid Replacement Derivatives

Decrease of oxytocic activity by aminoacid replacements in oxytocine of the N=42 compounds of Table 14 was studied also in multiparametric correlation [14,244]. The other structural parameters taken into account (Table 18) were the sum of differences, produced by all aminoacidic replacements for Tanford hydrophobicities ($\Delta\pi$), aromaticity character (ΔAR), charge transfer character (ΔCT), electric charge at pH7, (ΔEC), hydrogen bonding character (ΔHB), and orientation of dipole moment ($\Delta\mu$). The values for these intermolecular force parameters are those listed in §2.5. Histidine is considered with positive charge at pH7. Partial correlation coefficients are $r(\pi)=0.487$, $r(AR)=0.590$, $r(CT)=0.210$, $r(EC)=4.07$, $r(HB)=0.503$, $r(\mu)=0.567$, $r(MSD)=0.545$. The regressional equation obtained is :

$$A = 0.993+0.484\Delta\pi+0.646\Delta(AR)+0.761\Delta(CT)-0.685\Delta(EC)$$

$$+ 0.182\Delta(HB)+0.958\Delta\mu +0.067\ MSD \qquad (3)$$

$$r = 0.740, \quad s = 1.185, \quad N = 42$$

It is interesting that neglect of MSD decreases insignificantly r, which may be due to the large intercorrelation coefficients of MSD, especially with $\Delta(AR)$, r=0.816, with $\Delta(EC)$, r=0.702, and $\Delta(HB)$, r=0.700. Correlation of ΔA with the six intermolecular force parameters (π, AR, CT, EC, HB, μ) for the N=21 derivatives form Table 14 with replacements only in positions 3, 4 and 8 gives r=0.950. This fact suggests that steric fit in these positions is less important. This further suggests to use in correlation the sum of MSD's produced by substitutions only in positions 2 and 5 ($MSD^{2,5}$) together with $\Delta\pi$ for all substitutions as above. The correlational equation thus obtained is :

Table 18. *Differences in structural parameters introduced by amino-acid replacement*

Substitution	$\Delta\pi$	ΔAR	ΔCT	ΔEC	ΔHB	$\Delta\mu$	MSD
Phe-Tyr	0.32	0	0	0	1	1	1
Ser-Tyr	2.93	1	0	0	1	1	6
His-Tyr	1.87	0.5	0	1	0.25	1	2
Phe-Ile	0.32	1	0	0	0	0	5
Tyr-Ile	0.00	1	0	0	1	1	6
Ser-Ile	2.93	0	0	0	0	0	2
Trp-Ile	0.03	1.5	1	0	1	1	2
Leu-Ile	0.48	0	0	0	0	0	0
Val-Ile	1.28	0	0	0	0	0	1
His-Ile	1.87	0.5	0	1	1.25	0	4
Ser-GlN	0.14	0	0	0	0.5	0	3
Ala-GlN	0.83	0	0	0	2	0	4
AsN-GlN	0.09	0	0	0	0	0	3
Ser-AsN	0.05	0	0	0	0.5	0	2
Ala-AsN	0.74	0	0	0	2	0	3
Val-AsN	1.70	0	0	0	2	0	3
Val-Leu	0.80	0	0	0	0	0	3
Lys-Leu	0.99	0	0	1	1.25	0	3
Arg-Leu	1.76	0	0	1	1.25	0	5
His-Leu	1.39	0.5	0	1	1.25	0	2

$$\hat{A} = 1.931 + 0.240\Delta\pi + 0.458 \; MSD^{2,5}; \quad N=42, \quad r=0.623,$$
$$s=1.376; \quad r(MSD^{2,5}) = 0.603, \quad r(\Delta\pi, MSD^{2,5}) = 0.597 \tag{4}$$

While the partial correlation coefficient $r(MSD^{2,5})$ is somewhat higher than $r(MSD)$, the decrease in r, as compared to the 7-parameter equation (3) is nevertheless significant. Anyhow, r=0.623 of Eq.(4) is better than for the equation with $\Delta\pi$ and the, total, MSD :

$$\hat{A} = 1.021 + 0.338\Delta\pi + 0.210 \; MSD; \quad N=42, \quad r=0.593 \tag{5}$$

One can conclude that, in the receptor, the space juxtaposed to the side chains of aminoacids in position 3,4 and 8 is of rather low rigidity so that changes in intermolecular force parameters explain about 90% of the experimental activity decrease - variance for the corresponding N=21 oxitocine derivatives. Steric fit is more important for the receptor space juxtaposed to side chains in positions 2 and 5, but $MSD^{2,5}$ and $\Delta\pi$ are not sufficient as parameters and/or, the side chains of Tyr and GlN, respectively, are not near enough to complementarity with the receptor cavity and thus, not very good standards.

4.2.2 Carbamates as Acetyl Cholinesterase Inhibitors

A large series of aryl (phenyl-substituted) methyl carbamates was studied as inhibitors of bovine acetylcholinesterase by O'Brien[225] and by Metcalf et al.[226] These compounds should act by combining with the enzyme EOH :

$$CH_3NHCOOC_6H_4X + EOH \rightarrow CH_3NHCOOE + HOC_6H_4X$$

Correlations of $A = -\log I_{50}$ (molar concentrations producing 50% inhibition of enzyme activity) with different structural parameters were studied by Chiriac et al.[227] These parameters include HB, EC, CT, and MSD, as remembered in the previous paragraph, and also molar refraction MR and Hammett σ-constants, all for the substituent(s) in the phenyl cycle. Tab.19 lists the substituents X, the activity A and the values for substituent (X) parameters.

The best regressional equation without MSD is :

$$\hat{A} = 4.200-0.768\sigma+0.058\ MR+1.082\ EC+0.440\ HB+0.320\ CT$$
$$r=0.764, \quad s=0.809 \tag{6}$$

while inclusions of MSD gives :

$$\hat{A}=5.161-0.540\sigma+0.057\ MR+0.638\ EC+0.476\ HB+0.058\ CT-0.160\ MSD$$
$$r=0.800, \quad s=0.768, \quad F=22, \quad EV=61\% \tag{7}$$

The partial correlation coefficients are : $r(\sigma)=-0.159$, $r(MR)=0.630$, $r(EC)=0.424$, $r(HB)=0.418$, $r(CT)=-0.285$, $r(MSD)=0.312$, $r(\pi)=-0.045$. Inclusions of hydrophobicity, π, does not improve sensibly the correlation. The negative regressional coefficient and partial correlation coefficient for σ-Hammett indicates that inhibition is favored by electron-repelling substituents. The standard for MSD calculations was the most active compound, the 3,5-di-isopropyl derivative. This suggests the following conformation of acetylcholine in the receptor of acetylcholinesterase :

Table 19. *Parametrization of phenyl-substituted phenyl-methyl cambamates*

No	Substituent	A	σ	MR	HB	EC	CT	MSD
1	m-CH_3	4.853	-0.07	5.65	0	0	0	5.0
2	m-iso-C_3H_7	6.468	-0.09	14.97	0	0	0	3.0
3	m-tert-C_4H_9	6.398	-0.10	19.42	0	0	0	4.0
4	3,5-di-CH_3	5.222	-0.14	11.31	0	0	0	4.0
5	3-CH_3, 5-iso-C_3H_7	7.252	-0.16	20.63	0	0	0	2.0
6	2-CH_3, 5-iso-C_3H_7	5.699	-0.26	20.63	0	0	0	4.0
7	3,5 di-iso-C_3H_7	7.481	-0.18	29.95	0	0	0	0.0
8	3,5 di-tert-C_4H_9	7.108	-0.20	38.84	0	0	0	2.0
9	o-iso-C_3H_7O	6.174	-0.39	16.74	-1	0	0	7.0
10	2-F	4.796	0.24	0.81	0	0	0	7.0
11	2-Cl	5.301	0.20	5.84	0	0	0	7.5
12	2-Br	5.657	0.21	8.74	0	0	0	8.0
13	2-I	6.097	0.21	13.95	0	0	0	8.0
14	4-Cl	3.620	0.28	5.84	0	0	0	7.5
15	3-$N(CH_3)_2$	5.097	-0.21	15.55	0	0	-1	3.0
16	3,5-di CH_3, 4-CH_3S	5.921	-0.05	28.87	0	0	-1	4.5
17	3,5-di CH_3, 4-CH_3SO	5.745	0.42	33.81	-1	0	0	5.5
18	3,5-di CH_3, 4-CH_3SO_2	4.678	0.59	38.75	-2	0	0	6.5
19	o-$\overset{+}{N}(CH_3)_3$	5.000	0.91	21.20	0	1	0	8.0
20	m-$\overset{+}{N}(CH_3)_3$	7.097	0.94	21.20	0	0	0	4.0
21	p-$N(CH_3)_2$	3.620	-0.60	15.55	0	0	-1	7.0
22	p-$\overset{+}{N}(CH_3)_2$	5.456	0.86	21.20	0	1	0	8.0
23	3-$N(CH_3)_2$, 5-iso-C_3H_7	6.721	-0.30	30.52	0	0	-1	0.0
24	3-$\overset{+}{N}(CH_3)_3$, 5-iso-C_3H_7	8.155	0.81	36.18	0	1	0	1.0
25	4-$N(CH_3)_2$, 3-iso-C_3H_7	6.824	-0.51	30.52	0	0	-1	4.0
26	4-$\overset{+}{N}(CH_3)_3$, 3-iso-C_3H_7	8.301	0.77	36.18	0	1	0	5.0
27	3,5-di-$N(CH_3)_2$	5.585	-0.42	31.10	1	1	-1	0.0
28	3-$N(CH_3)_2$, 5-$N(CH_3)_2$	7.432	0.69	36.75	0	1	-1	1.0
29	3,5-di-$\overset{+}{N}(CH_3)_3$	6.921	1.81	42.40	0	2	1	2.0
30	4-$N(CH_3)_2$, 6-CH_3, 3-iso-C_3H_7	6.409	-0.87	35.18	0	0	-1	5.0
31	4-$\overset{+}{N}(CH_3)_3$, 6-CH_3, 3-iso-C_3H_7	7.959	0.60	40.83	0	1	0	6.0
32	4-$N(CH_3)_2$, 5-CH_3, 2-iso-C_3H_7	5.886	-1.07	35.18	0	0	-1	7.0
33	4-$\overset{+}{N}(CH_3)_3$, 5-CH_3, 2-iso-C_3H_7	8.000	0.39	40.83	0	1	0	8.0

Table 19 (continued)

No	Substituent	A	σ	MR	HB	EC	CT	MSD
34	m-CH_3S	5.155	0.23	16.56	0	0	-1	3.5
35	m-$(CH_3)_2S^+$	6.187	1.03	35.13	0	1	0	3.5
36	p-CH_3S	4.468	0.22	16.56	0	0	-1	6.5
37	p-$(CH_3)_2S^+$	4.958	1.20	35.13	0	1	0	5.5
38	m-$(C_2H_5)_2P$	6.131	0.10	38.88	0	0	-1	5.5
39	m-$(C_2H_5)_2P(CH_3)$	7.444	0.80	44.53	0	1	0	6.5
40	H	3.699	0.00	1.10	0	0	0	6.0
41	o-CH_3	3.853	-0.17	5.65	0	0	0	7.0
42	p-CH_3	4.000	-0.17	5.65	0	0	0	7.0
43	o-C_2H_5	4.886	-0.23	10.30	0	0	0	6.0
44	m-C_2H_5	5.319	-0.04	10.30	0	0	0	4.0
45	p-C_2H_5	4.420	-0.15	10.30	0	0	0	6.0
46	o-iso-C_3H_7	5.222	-0.30	14.97	0	0	0	7.0
47	p-iso-C_3H_7	4.155	-0.15	14.97	0	0	0	5.0
48	o-tert-C_4H_9	5.222	-0.64	19.42	0	0	0	8.0
49	p-tert-C_4H_9	5.824	-0.20	19.42	0	0	0	6.0
50	o-sec-C_4H_9	5.959	-0.59	19.62	0	0	0	8.0
51	m-sec-C_4H_9	6.796	-0.11	19.62	0	0	0	4.0
52	p-sec-C_4H_9	5.745	-0.12	19.62	0	0	0	6.0
53	m-sec-C_5H_{11}	6.959	-0.22	24.20	0	0	0	5.0
54	o-cyclo-C_5H_9	5.959	-0.91	24.56	0	0	0	9.0
55	m-cyclo-C_5H_9	5.824	-0.03	24.56	0	0	0	5.0
56	p-cyclo-C_5H_9	4.569	-0.03	24.56	0	0	0	7.0
57	o-cyclo-C_6H_{11}	5.854	-1.19	26.20	0	0	0	10.0
58	m-cyclo-C_6H_{11}	5.699	-0.02	26.20	0	0	0	6.0
59	p-cyclo-C_6H_{11}	5.046	-0.02	26.20	0	0	0	8.0
60	o-C_3H_7	5.268	-0.56	14.96	0	0	0	7.0
61	o-iso-C_4H_9	5.638	-1.13	19.62	0	0	0	8.0
62	m-F	4.071	0.34	0.81	0	0	0	5.0
63	p-F	3.638	0.06	0.81	0	0	0	7.0
64	m-Cl	4.301	0.37	5.84	0	0	0	4.5
65	m-Br	4.886	0.39	8.74	0	0	0	4.0
66	p-Br	4.000	0.23	8.74	0	0	0	8.0
67	m-I	5.155	0.35	13.95	0	0	0	4.0
68	p-I	4.055	0.28	13.95	0	0	0	8.0
69	2,3-di-Cl	4.319	0.57	11.69	0	0	1	6.0
70	2,4-di-Cl	4.853	0.43	11.69	0	0	1	9.0

Table 19 (continued)

No	Substituent	A	σ	MR	HB	EC	CT	MSD
71	2,5-di-Cl	4.301	0.57	11.69	0	0	1	6.0
72	2,6-di-Cl	2.886	0.40	11.69	0	0	1	9.0
73	3,4-di-Cl	4.723	0.60	11.69	0	0	1	6.0
74	3,5-di-Cl	4.921	0.75	11.69	0	0	1	3.0
75	o-NO$_2$	2.301	1.22	6.71	-2	0	1	7.0
76	m-NO$_2$	2.699	0.71	6.71	-2	0	1	3.0
77	p-NO$_2$	2.523	0.78	6.71	-2	0	1	5.0
78	2-NO$_2$, 3-CH$_3$	3.699	1.15	12.37	-2	0	1	8.0
79	2-NO$_2$, 4-CH$_3$	3.886	1.05	12.36	-2	0	1	8.0
80	2-NO$_2$, 5-CH$_3$	4.796	0.54	12.37	-2	0	1	6.0
81	3-NO$_2$, 4-CH$_3$	3.495	0.71	12.37	-2	0	1	4.0
82	4-NO$_2$, 3-C$_2$H$_5$	3.699	0.74	17.04	-2	0	1	5.0
83	4-NO$_2$, 3-iso-C$_3$H$_7$	5.553	0.69	21.70	-2	0	1	4.0
84	p-C$_2$H$_5$S	4.251	0.04	22.21	0	0	-1	7.5
85	p-C$_3$H$_7$S	4.921	0.06	26.87	0	0	-1	8.5
86	p-iso-C$_3$H$_7$S	5.046	0.04	26.88	0	0	-1	8.5
87	p-C$_4$H$_9$S	5.432	0.04	31.50	0	0	-1	7.5
88	p-C$_6$H$_{13}$S	5.398	0.39	40.76	0	0	-1	9.5
89	m-C$_8$H$_{17}$S	5.921	0.40	50.05	0	0	-1	10.5
90	p-C$_8$H$_{17}$S	4.886	0.48	50.05	0	0	-1	11.5
91	p-CH$_3$SO	4.796	0.57	22.50	-1	0	0	5.5
92	p-CH$_3$SO$_2$	4.000	0.73	27.44	-2	0	0	6.5
93	4-CH$_3$S, 3-iso-C$_3$H$_7$	7.000	0.13	31.54	0	0	-1	3.5
94	6-CH$_3$S, 3-iso-C$_3$H$_7$	6.745	-0.26	20.63	0	0	-1	3.5
95	m-CH$_2$=CHCH$_2$S	5.444	0.14	36.22	0	0	-1	5.5
96	p-CH$_2$=CHCH$_2$S	5.067	0.13	36.22	0	0	-1	7.5
97	p-CH≡CCH$_2$S	4.770	0.13	24.10	0	0	-1	7.5

$$CH_3CO-O-CH_2$$
$$\backslash$$
$$CH_2-\overset{+}{N}(CH_3)_3$$

similar to

$$CH_3NHCO-O-\bigcirc-\overset{+}{N}(CH_3)_3$$

Suggested conformation of acetylcholine in the receptor of acetyl-cholinesterase.

4.2.3 Schrader-type Organophosphorus Compounds

Organic derivatives of phosphoric and thiophosphoric acids are largely used as pesticides although they also present high toxicity against mammals. Their toxic action is due to acetylcholinesterase inhibition owing to alkylation of the enzyme EOH[228] :

$$R_1 \underset{R_2}{\overset{}{\diagup}} P \overset{O}{\diagdown}_{OR_3} \quad + \quad EOH \quad \longrightarrow \quad R_1 \underset{R_2}{\overset{}{\diagup}} P \overset{O}{\diagdown}_{OE} \quad + \quad R_3OH$$

Structure activity correlations, with MSD, for a large series of this type of compounds were studied by Vilceanu, Chiriac et al.[221,229,230] The series of N=71 compounds are Schrader compounds of the general type :

$$R_1 \underset{R_2}{\overset{}{\diagup}} P \overset{O}{\diagdown}_{OR_3}$$

R_1, R_2 are alkoxy groups, $-OC_nH_{2m+1}$, but also β-chloroethyl groups and for some compounds, groups like CH_3, CCl_3 and CH_3S replace one alkoxy substituent. R_3 is a relatively acidic group as : $-CH:CCl_2$, $-C(C_6H_5):CH_2$, $-C(CH_3):CHCOOC_2H_5$, $-P(O)(OCH_3)_2$, $4-NO_2-C_6H_4-$, etc. The biological activities used are A= -log LD_{50} (lethal doses, 50% for mice). The following structural parameters are used: σ_{12}^ϕ constants for MR_1+MR_2, n_{12}-number of atoms (Y:) with nonbonded electron pairs (H-bonding) also for R_1+R_2, molar refractivity MR_{12} for R_1+R_2, σ_3^ϕ-for R_3, n_3 for R_3, AR-number of phenylic cycles in R_3, CT-number of electron-attracting groups (NO_2 or CN) in R_3, electron attracting in charge transfer, MR_3-refractivity of R_3 and MSD. The standard for MSD calculations is the natural effector, acetylcholine S_A :

$$S_A : \quad \ldots -CH_2 \underset{}{\overset{O}{\diagup}} C \diagdown_{O-CH_2-CH_2-N^+ \underset{CH_3}{\overset{CH_3}{\diagup}} CH_2} - \ldots$$

but the groups (atoms) which stick along the free valencies are considered to stick out into the environment and to be therefore irrelevant for steric fit; they are not counted in calculating MSD. Thus R_1 and R_2 do not contribute to MSD (if at least one of them is as large as CH_3), also atoms of linear chains as R_3 after the 4-th term.

Correlations were performed separately for the N=71 compounds and all k=9 parameters or with k=6 parameters $(\sigma_{12}^{\phi} + \sigma_{3}^{\phi}, n_{12}+n_3$ and $R_{12}+R_3$ as single parameters) and also for a reduced series of N=51 compounds, in which the compounds without a markedly acidic R_3-group are excluded. Also a correlation including quadratic terms (except MSD) i.e., with k=17 was performed. Results of these correlations are listed in Table 20. The best variant is (c), with the highest Fischer-test value F=13.8. The corresponding correlational equations is :

Table 20. Results of correlations

Variant	Eq.	N	k	r	s	F
a	-	51	6	0.658	0.758	4.7
b	-	51	9	0.663	0.754	3.3
c	(8)	71	6	0.716	0.746	13.8
d	-	71	9	0.713	0.749	7.8
e	-	71	17	0.801	-	4.7

$$\hat{A} = 0.555 + 0.476(\sigma_{12}^{\phi} + \sigma_{3}^{\phi}) - 0.017(n_{12}+n_3) + 0.343(MR_{12}+MR_3) -$$
$$-0.402\ AR + 0.611\ CT - 0.063\ MSD \tag{8}$$
$$n = 71, \quad r = 0.716$$

The partial correlation coefficients corresponding to N=71, k=6 (Eq.8) are : $r(\sigma_{12}^{\phi}+\sigma_3)=0.508$; $r(n_{12}+n_3)=-0.240$; $r(MR_{12}+MR_3)=0.289$; $r(AR)=-0.078$; $r(CT)=0.285$; $r(MSD)=0.048$. The low partial correlation coefficient with MSD indicates that acetylcholine is not a good standard, although compounds with R_3 of the type

similar to

representing a moiety with a possible acetylcholine conformation, are among the most active ones. The rather high partial correlation coefficient $r(\sigma_{12}^{\phi}+\sigma_3^{\phi})$ suggests that enzyme acylation is always the mechanism of inhibition for the $R_1R_2P(O)OR_3$-compounds. Anyhow, a large number of N=71 compounds with largely different structures could be described by the statistically significant regressional Equation (8).

The lack of correlation of A vs. MSD to acetylcholine suggests that these inhibitors bind in different ways to the enzyme and that different standards should be used for different subseries of compounds. The compounds with ethylenic R_3 of Table 21 did not yield a good unique correlation equation, but good correlations could be found for the series of phenyl substituted (at C_α of the ethylenic group) compounds and for the last, methyl substituted compounds.[230,231,232].

Table 21. Organophosphoric Schrader-type compounds $R_1 R_2 0(0) OR_3$ with ethylenic R_3-group.

Nr	R_1, R_2, R_3	A_i	σ_3^ϕ	$MR_{12} + MR_3$	MSD
1	$-OCH_3, -OCH_3, -C(C_6H_5) = CClCOOC_2H_5$	1.56	4.07	69	0
2	$-OCH_3, -OCH_3, -C(C_6H_5) = CHCOOC_2H_5$	1.40	1.49	54	1.5
3	$-OC_2H_5, -OC_2H_5, C(C_6H_5) = CHOC_6H_4NO_2(p)$	1.37	1.55	90	2
4	$-OC_2H_5, -OC_2H_5, -C(C_6H_5) = CClOC_6H_4NO_2(p)$	1.35	4.07	78	2
5	$-OCH_3, -OCH_3, -C(C_6H_5) = CClOC_6H_4NO_2(p)$	1.30	3.27	85	2
6	$-OCH_3, -OCH_3, -C(C_6H_5) = CHOC_6H_4NO_2(p)$	0.94	1.55	80	3
7	$-OC_2H_5, -OC_2H_5, -C(C_6H_5) = CHCOOC_2H_5$	0.48	1.43	74	4
8	$-OC_2H_5, OC_2H_5, -C(C_6H_5) = CHCl$	0.94	1.24	63	6
9	$-OCH_3, -OCH_3, -C(C_6H_5) = CHC_6H_5$	-0.52	0.33	73	9
10	$-OC_2H_5, -OC_2H_5, -C(C_6H_5) = CH_2$	-0.59	0.81	58	7.5
11	$-OCH_3, -OCH_3, -C(CH_3) = C(CN)C_6H_5$	-0.17	3.08	58	10
12	$-OC_2H_5, -OC_2H_5, -C(CH_3) = C(CN)C_6H_5$	0.13	3.08	67	8
13	$-OCH_3, -OCH_3, -C(CH_3) = C(COOEt)CH_2COOEt$	0.41	2.16	59	7
14	$-OC_2H_5, -OC_2H_5, -C(CH_3) = C(C_6H_5)COOC_2H_5$	0.45	1.98	78	6
15	$-OCH_3, -OCH_3, -C(CH_3) = CHCOCH_3$	0.58	0.72	38	4
16	$-OC_2H_5, -OCH_3, -C(CH_3) = CHCOCH_3$	0.99	0.73	48	2
17	$-OC_2H_5, -OC_2H_5, -C(CH_3) = CClCOOC_2H_5$	1.58	3.57	59	1

For the N=10 phenyl substituted compounds, standard S_{Ph} was used, for the N=7 methyl substituted compounds, standard S_{Me} :

S_{Me} :

$$CH_3CH_2O \diagdown P \diagup {}^{=O}$$

$$CH_3CH_2O \diagup P \diagdown O-C=CH-CO-OEt$$

$$| \ CH_3$$

S_{Ph} :

$$CH_3O \diagdown P \diagup {}^{=O}$$

$$CH_3O \diagup P \diagdown O-C=CCl-CO-OEt$$

$$| \ C_6H_5$$

For the α-methyl substituted compounds the best uni- and biparametric correlations are :

$$\hat{A}=1.472-0.167 \ MSD, \ N=10, \ r=0.964 \tag{9a}$$

$$\hat{A}=1.244+0.116 \ \sigma_3^\phi - 0.175 \ MSD, \ N=10, \ r=0.987, \ s=0.091 \tag{9b}$$

while for the α-phenyl substituted compounds :

$$\hat{A}=1.766-0.271 \ MSD, \ N=7, \ r=0.971 \tag{10a}$$

$$\hat{A}=1.137+0.006(MR_{12}+MR_3)-0.173 \ MSD, \ N=8, \ r=0.974, \ s=0.136 \tag{10b}$$

Acethylcholine came out to be a good MSD standard for acetyl-cholin-esterase phosphorylation by Schrader compounds of the type $R_1R_2=P(O)R_3$, with R_3 as cationic group[221], listed in Table 22. A is the logarithm of the bimolecular rate constant for the phosphory-lation of the enzyme. The standard used for MSD calculations was S_A depicted at the beginning of this paragraph. The following regres-sional equation was obtained.

$$\hat{A}=2.941+0.136\pi-0.060\sigma^\phi-0.028 \ CT-0.272 \ MSD+3.003 \ EC \tag{11}$$

$$N=60, \ r=0.880, \ s=0.906$$

The structural parameters refer to the sum of parameters for R_1,R_2 and R_3. The partial correlation coefficients and intercorrelation coeffi-cients are listed in Table 23. Thus Schrader type compounds with ca-tionic groups $-S-CH_2CH_2\overset{+}{S}R_2$ or $-S-CH_2CH_2\overset{+}{N}R_3$ probably bind in the same positions to acetylcholinesterase as acetylcholine.

Other short series of acetyl- and butyrylcholinesterase inhi-bitors are listed in Table 24. The rather satisfactory correlation ob-tained with different standards for MSD suggest that different classes of inhibitors bind to different sites on the enzyme[231,232].

Table 22. Schrader compounds $R_1R_2P(O)R_3$ (correlation Equation (11))

Nr.	R$_1$	R$_2$	R$_3$	A	MSD	EC
1	CH_3	$(CH_3)_2CHO$	$SCH_2CH_2CH(CH_3)_2$	1.08	1.5	0
2	CH_3	$(CH_3)_2CHO$	$SCH_2CH_2SCH_3$	2.40	2.0	0
3	CH_3	$(CH_3)_2CHO$	$SCH_2CH_2SC_2H_5$	2.55	2.0	0
4	CH_3	$(CH_3)_2CHO$	SC_2H_5	0.97	4.5	0
5	CH_3	$(CH_3)_2CHO$	SCH_2CH_2F	2.08	3.5	0
6	CH_3	$(CH_3)_2CHO$	$SCH_2CH_2S^+(CH_3)_2$	5.80	1.0	1
7	CH_3	$(CH_3)_2CHO$	$SCH_2CH_2NH^+(CH_3)_2$	5.23	1.5	1
8	CH_3	$(CH_3)_2CHO$	$SCH_2CH_2N^+(CH_3)_3$	5.95	0.5	1
9	CH_3	$(CH_3)_2CHO$	F	5.36	6.0	0
10	CH_3	$(CH_3)_2CHO$	$O-C_6H_4-NO_2-p$	4.25	4.0	0
11	CH_3	$(CH_3)_2CHO$	$O-C_6H_4-N(CH_3)_2-m$	1.83	4.0	0
12	CH_3	C_2H_5O	$SCH_2CH_2SCH_3$	2.83	2.0	0
13	CH_3	C_2H_5O	$SCH_2CH_2SC_2H_5$	3.00	2.0	0
14	CH_3	C_2H_5O	$SCH_2CH_2SC_4H_9-n$	3.17	2.0	0
15	CH_3	C_2H_5O	$SCH_2CH_2SC_6H_{13}-n$	3.83	2.0	0
16	CH_3	C_2H_6O	$SCH_2CH_2SC_8H_{17}-n$	3.74	2.0	0
17	CH_3	C_2H_5O	$SCH_2CH_2SC_{10}H_{21}-n$	3.52	2.0	0
18	CH_3	C_2H_5O	$SCH_2CH_2S^+(CH_3)CH_3$	5.99	0.5	1
19	CH_3	C_2H_5U	$SCH_2CH_2S^+(CH_3)C_2H_5$	6.63	0.5	1
20	CH_3	C_2H_5O	$SCH_2CH_2S^+(CH_3)C_4H_9-n$	6.57	0.5	1
21	CH_3	C_2H_5O	$SCH_2CH_2S^+(CH_3)C_6H_{13}-n$	7.28	0.5	1
22	CH_3	C_2H_5O	$SCH_2CH_2S^+(CH_3)C_8H_{17}-n$	6.86	0.5	1
23	CH_3	C_2H_5O	$SCH_2CH_2S^+(CH_3)C_{10}H_{21}-n$	6.83	0.5	1
24	CH_3	CH_3O	SCH_3	0.63	5.5	0
25	CH_3	C_2H_5O	SCH_3	1.30	5.5	0
26	CH_3	$n-C_3H_7O$	SCH_3	1.97	5.5	0
27	CH_3	$n-C_4H_9O$	SCH_3	1.36	5.5	0
28	CH_3	$n-C_5H_{11}O$	SCH_3	1.04	5.5	0
29	CH_3	$n-C_6H_{13}O$	SCH_3	0.85	5.5	0
30	CH_3	$n-C_7H_{15}O$	SCH_3	0.87	5.5	0
31	CH_3	$n-C_8H_{17}O$	SCH_3	0.84	5.5	0
32	CH_3	$n-C_9H_{19}O$	SCH_3	1.00	5.5	0
33	CH_3	$n-C_{10}H_{21}O$	SCH_3	0.68	5.5	0
34	CH_3	C_2H_5O	$SCH_2CH_2OC_6H_6$	2.62	2.5	0
35	CH_3	C_2H_5O	$SCH_2CH_2OC_6H_4CH_3-m$	2.52	2.5	0
36	CH_3	C_2H_5O	$SCH_2CH_2OC_6H_4CH_3-p$	2.45	2.5	0

Table 22 (continued)

Nr.	R_1	R_2	R_3	A	MSD	EC
37	CH_3	C_2H_5O	$SCH_2CH_2OC_6H_4CHO-m$	2.83	2.5	0
38	CH_3	C_2H_5O	$SCH_2CH_2OC_6H_4CHO-p$	2.68	2.5	0
39	CH_3	C_2H_5O	$SCH_2CH_2OC_6H_4Cl-m$	3.15	2.5	0
40	CH_3	C_2H_5O	$SCH_2CH_2OC_6H_4Cl-p$	3.11	2.5	0
41	CH_3	C_2H_5O	$SCH_2CH_2OC_6H_4Br-p$	3.05	2.5	0
42	CH_3	C_2H_5O	$SCH_2CH_2OC_6H_4(tC_4H_9)-p$	3.43	2.5	0
43	CH_3	C_2H_5O	$SCH_2CH_2OC_6H_4NO_2-m$	3.95	2.5	0
44	CH_3	C_2H_5O	$SCH_2CH_2OC_6H_4NO_2-p$	4.04	2.5	0
45	CH_3	C_2H_5O	$SCH_2CH_2OC_6H_4OH-m$	3.23	2.5	0
46	CH_3	C_2H_5O	$SCH_2CH_2OC_6H_4OH-p$	3.96	2.6	0
47	C_2H_5O	C_2H_5O	$SCH_2CH_2N(CH_3)C_6H_4OCH_3-p$	2.92	1.5	0
48	C_2H_5O	C_2H_5O	$SCH_2CH_2N(CH_3)C_6H_4CH_3-p$	2.49	1.5	0
49	C_2H_5O	C_2H_5O	$SCH_2CH_2N(CH_3)C_6H_4CH_3-m$	2.20	1.5	0
50	C_2H_5O	C_2H_5O	$SCH_2CH_2N(CH_3)C_6H_5$	2.04	1.5	0
51	C_2H_5O	C_2H_5O	$SCH_2CH_2N(CH_3)C_6H_4OCH_3-m$	2.20	1.5	0
52	C_2H_5O	C_2H_5O	$SCH_2CH_2N(CH_3)C_6H_4Cl-p$	2.04	1.5	0
53	C_2H_5O	C_2H_5O	$SCH_2CH_2N(CH_3)C_6H_4Cl-m$	1.52	1.5	0
54	C_2H_5O	C_2H_5O	$SCH_2CH_2N^+(CH_3)_2C_6H_4OCH_3-p$	5.68	0.5	1
55	C_2H_5O	C_2H_5O	$SCH_2CH_2N^+(CH_3)_2C_6H_4CH_3-p$	5.60	0.5	1
56	C_2H_5O	C_2H_5O	$SCH_2CH_2N^+(CH_3)_2C_6H_4CH_3-m$	5.72	0.5	1
57	C_2H_5O	C_2H_5O	$SCH_2CH_2N^+(CH_3)_2C_6H_5$	5.62	0.5	1
58	C_2H_5O	C_2H_5O	$SCH_2CH_2N^+(CH_3)_2C_6H_4OCH_3-m$	5.85	0.5	1
59	C_2H_5O	C_2H_5O	$SCH_2CH_2N^+(CH_3)_2C_6H_4Cl-p$	5.57	0.5	1
60	C_2H_5O	C_2H_5O	$SCH_2CH_2N^+(CH_3)_2C_6H_4Cl-m$	5.81	0.5	1

Table 23. Partial correlation coefficients and intercorrelation coefficients (Eq.11)

	log k	σ	π	MSD	EC	CT
σ	0.205	1.000	0.361	-0.113	0.141	0.325
π	0.296		1.000	0.045	0.465	0.503
MSD	-0.717			1.000	-0.645	0.080
EC	0.852				1.000	0.301
CT	0.189					1.000

Table 24. Other acetyl and butyrilcholinesterase inhibitors

Series of compounds	Standard	N	Results
$\overset{O}{\overset{\|}{EtO-PMe}}-S(CH_2)_nR$	$\overset{O}{\overset{\|}{EtO-PMe}}-SCH_2-CMe_3$	24	$A(\pi,MSD)$, r=0.920 $B(\pi,MSD)$, r=0.852
$\overset{O}{\overset{\|}{RO-PMe}}-SC_4H_9$	$n-C_4H_9-\overset{O}{\overset{\|}{PMe}}-SC_4H_9$	10	$A(\pi,MSD)$, r=0.981
$\overset{O}{\overset{\|}{RO-PMe}}-S(CH_2)SEt$	$PhO-\overset{O}{\overset{\|}{PMe}}-S(CH_2)_2SEt$	13	$A(\pi,MSD)$, r=0.960
$\overset{O}{\overset{\|}{RO-PMe}}-S(CH_2)_2\overset{+}{S}MeEt$	$n-C_8H_{17}O-\overset{O}{\overset{\|}{PMe}}-S(CH_2)_2\overset{+}{S}MeEt$	8	$A(MR,MSD)$,r=0.945 $A(\sigma,MSD)$, r=0.975 $A(E_S,MSD)$,r=0.992
$\overset{O}{\overset{\|}{Ph_2P}}-SR$	$\overset{O}{\overset{\|}{Ph_2P}}-S-C_6H_{13}-n$	9	$B(\pi,MSD)$, r=0.992 $B(E_S,MSD)$,r=0.989
$Et_2N-\overset{O}{\overset{\|}{PMe}}-NHC_6H_4R$	$Et_2N-\overset{O}{\overset{\|}{PMe}}-NHC_6H_4NO_2p$	10	$A(\sigma,MSD)$, r=0.782
$R_2NH-\overset{S}{\overset{\|}{P}}(OR_1)-SC_6H_4R$	$Me_3CNH-\overset{S}{\overset{\|}{P}}(OEt)-SPh$	12	$A(E_S,MSD)$, r=0.843

Notations: A(R,MSD), r=0.920 - means correlational equation for
acetylcholinesterase inhibition, with R and MSD as structural parameters, yields r=0.920. Correspondingly, B stand for butyrilcholinesterase. N - number of compounds in the series.

4.2.4 Potential Cytostatic Pt(II)-Diammines

cis-Dichlorodiammine Pt(II)-complexes have been tested on a large scale as potential cytostatic agents[233]. The large steric variety of ligands used, like diammine, ethyleneimine, ethylenediamine, cyclic aliphatic amines, aromatic amines, saturated and aromatic N-heterocycles suggested to attempt correlations of biological activity with structural parameters including MSD[234]. For MSD calculations the standards were obtained by trial and error, starting from the most active ammine. Lethal doses 50% for mice (LD_{50}), doses of 90% efficiency (ED_{90}) and therapeutic indices (TI) for various tumors were considered. A brief account of results follows.

For N=17 complexes with monodentate ligands, the correlation with MSD, pK and log P_{oct} (i.e., logarithms of octanol: water partition coefficients) of the ammines gave r=0.90, with partial correlation coefficients r(logP_{oct})=0.89, r(pK)=0.08 and r(MSD)= -0.71, for LD_{50}. For ED_{50}, r=0.68 was obtained in the triparametric correlation with r(logP_{oct})= 0.62, r(pK)= -0.32 and r(MSD)=0.58. For TI, r=0.86 (triparametric), r(logP_{oct})=0.40, r(pK)=0.54 and r(MSD)= -0.79.

For N=28 complexes with mono- and bidentate ligands for the same tumors (ASD/PC6 on mice), for TI the triparametric correlation yields r=0.71, with r(logP_{oct})=0.37, r(pK)=0.31 and r(MSD)= -0.56.

For N=21 complexes tested against rat Sarcome 180, for TI, triparametric correlation gives r=0.63, with r(logP_{oct})=0.53, r(pK)=-0.53, r(MSD)=0.48.

The results suggest that correlations with MSD, using methods to optimize the standard may be useful in improving the therapeutic index for this type of cytostatics.

4.3 Comparison Between Topological Indices, MSD and Other Sterical Parameters

As already mentioned, the steric parameter $\bar{\Delta}$ of Amoore[8,9] is also a measure of shape similarity/difference between a molecule and a standard, like MSD. A comparison between $\bar{\Delta}$, the connectivity index χ of Randić and MSD will now be described[235].

The series studied is listed in Table 25, the first 30 compounds from the work of Amoore and Venstrom[81]. The odor similarity of these compounds relative to 1,2-dichlorethane - the standard for the fundamental ethereal odor, is considered as biological activity, A. Table 25 contains also the $\bar{\Delta}$ and MSD figures with respect to 1,2-dichlorethane and Randić's χ values for the 30 compounds. A and $\bar{\Delta}$ figures are taken from the work of Amoore and Venstrom[81].

The following regressional equations were obtained for this series of N=30 compounds :

$$\hat{A} = -5.486 + 12.834\bar{\Delta} \; , \quad r=0.62, \quad s=1.25, \quad F=8.33 \tag{12}$$

$$\hat{A} = 4.556 - 0.469 MSD; \quad r=0.77, \quad s=1.00, \quad F=20.25 \tag{13}$$

$$\hat{A} = 6.133 - 1.248\chi \; ; \quad r=0.73, \quad s=1.09, \quad F=15.10 \tag{14}$$

$$\hat{A} = 9.584 - 3.73\chi + 0.41\chi^2; \quad r=0.77, s=1.01, \quad F=12.47 \tag{15}$$

The stability of equations (12-15) was studied by extending the series with another 10 compounds (p-dichlorobenzene, A=1.69; 1,1,1,2,2,3,3-heptachloropropane, 1.31; hexachlorethane, 1.14; hexamethylethane, 1,26; isoquinoline, 0.86; methyl cyclopropyl ketone, 3.77; methylene-chloride, 6.27; naphthalene, 1.77; nitrobenzene, 1.26; n-octane, 1.16). For the N=30+10=40 compounds, the A vs $\bar{\Delta}$ correlation yields r=0.64, A vs χ:r=0.49; A vs χ and χ^2:r=0.79; and for the last N=10 compounds, A vs MSD yields r=0.82. The stability of Eqns(1-4) is reasonable.

The $\bar{\Delta}$-MSD intercorrelation coefficient is rather low, r=-0.65. This may be partly due to the fact that $\bar{\Delta}$ used for QSAR by Amoore et al. is a similarity index. $\bar{\Delta}=1/(1+\Delta_m)$, where Δ_m is directly related to the shape differences of molecular contours and thus Δ_m should be compared to MSD.

Table 25 Correlation of similarity to etheral odor Δ, MSD and Randic's topological index χ

Nr.	Compound	A	$\bar{\Delta}$	MSD	χ
1	acetone	4.42	0.687	2	1.353
2	acetophenone	0.71	0.584	5	3.013
3	1,1,2,2-tetrachloroethane	4.32	0.647	2	2.641
4	anisic aldehyde	0.77	0.579	6	3.171
5	anisole	2.65	0.630	6	2.970
6	benzaldehyde	1.27	0.643	4	2.607
7	benzene	3.62	0.639	2	1.998
8	benzonitrile	1.14	0.640	4	2.449
9	benzophenone	1.05	0.505	10	3.969
10	benzyl acetate	1.65	0.623	7	2.970
11	n-butyl propionate	1.32	0.599	5	3.473
12	chlorobenzene	2.16	0.668	3	2.490

Table 25 (continued)

Nr.	Compound	A	$\bar{\Delta}$	MSD	χ
13	chloroform	6.69	0.645	2	1.731
14	cyclohexane	3.66	0.609	2	3.000
15	cyclohexanol	1.51	0.618	3	3.393
16	cyclohexanone	1.91	0.627	3	3.059
17	cyclooctanone	1.55	0.545	5	3.559
18	cyclopentanone	2.33	0.698	2	2.559
19	cyclopentylacetate	0.91	0.580	5	3.577
20	di-n-butyl ether	1.29	0.582	5	4.414
21	cis-1,2-dichlorethylene	5.45	0.672	0	1.487
22	trans-1,2-dichlorethylene	3.41	0.793	2	1.487
23	diethyl ether	5.12	0.725	1	2.414
24	diethyl sulfate	1.86	0.531	5	3.570
25	dimethylbenzylcarbinol	1.02	0.504	7	3.660
26	dimethylbenzylcarbinylacetate	1.06	0.457	10	4.906
27	2,2-dimethylpropylacetate	1.34	0.546	5	3.745
28	diphenylether	1.14	0.503	9	4.526
29	ethyl acetate	3.48	0.698	2	2.600
30	eugenol	1.28	0.498	2	4.468

MSD values were calculated with respect to 1,2-dichlorethane in a conformation similar to that of cis-1,2-dichlorethylene.No difference between second period and higher period atoms was considered.

A comparative study of some steric parameters, including MSD, was performed by Verloop and Tipker[4] for four series :

Binding of $XC_6H_4OCOCH_2NHSO_2CH_3$ to papain (log $1/K_M$), N=13 derivatives was correlated with following combinations of parameters : σ and MR-yielding r=0.935; σ and χ_v-r=0.938; σ and u-r=0.878; σ and MSD -r=0.924 and σ and B_4-r=0.918. For the MSD procedure the standard is the derivative X=p-NO_2 with highest K_M. Here u is the normalized minimum width (Charton)[236].

Inhibition of bovine acetylcholinesterase by N=29 3,5-substituted phenyl-N-methylcarbamates (-log I_{50}) was correlated with following combinations of parameters : $\pi^{(3)}$, σ and EC-r=0.858; $\pi^{(3)}$, σ, EC, $MR^{(3)}$ and $MR^{(5)}$ -r=0.889; $\pi^{(3)}$, σ, EC and $MSD^{(3+5)}$-r=0.933; $\pi^{(3)}$, σ, EC, $B_1^{(3)}$ and $B_4^{(5)}$-r=0.940; $\pi^{(3)}$, σ, EC, $u^{(3)}$ and $u^{(5)}$-r=0.933; $\pi^{(3)}$, σ, EC, $\chi_v^{(3)}$ and $\chi_v^{(5)}$-r=0.899. Standard for MSD procedure was the

3,5-diisopropyl-substituted derivative with highest $-\log I_{50}$. EC is electric charge at pH7.

Larvicidal activities of N=26 p-substituted(phenyl)2,6 dichlorobenzoylureas to Pieris brassicae L. $(-\log LD_{50})$ was correlated with following parameter-combinations : π and σ:r=0.148; π, σ, L and B_4:r=0.848; π, σ and χ_v:r=0.701; π, σ and MSD:r=0.697; π, σ and MR : r=0.659; π, σ and u:r=0.638. Standard for MSD: the CF_3 substituted derivative.

Hapten antibody interaction of N=36 ortho-, meta- and para-substituted benzoic acids (K relative to K of unsubstituted compound), was correlated with following parameter combinations : B_1^0 and B_1^P:r=0.953; χ_v^0 and χ_v^P:r=0.723; MR^0 and MR^P:r=0.714; MSD^0 and MSD^P : r=0.655. Standard for MSD: the p-NHCOCH$_3$ substituted derivative. In this study, the STERIMOL-parameters $L, B_1 - B_4$ give the best results, which might be generally true if the molecular shape differences are produced by various, not too large, substituents on a fixed molecular core.

5. MTD-RECEPTOR SITE MAPPING

The original idea of minimal steric difference is that affinity towards a receptor decreases linearly with the nonoverlappable volumes of the considered molecule and the receptor[14]. The point is that the shape of the receptor cavity is not known and also that parts of the molecules stick out into the aqueous environment, being irrelevant for steric fit[20]. Consider the situation depicted in Figure 9 with a receptor cavity that accommodates exactly a benzenic cycle, and a set of molecules with determined biological activities, A_i consisting of phenol(S), cyclohexanone(I), t-butylmethylether(II) and n-hexylethylether(III). According to the MSD-procedure, phenol must be considered as standard(S) and we shall search a maximal superposition of the other molecules upon it. The result of these superpositions is a topological network, the hypermolecule $\hat{H}$, which here has 11 vertices, corresponding to the approximate positions of atomic (nonhydrogen) nuclei. Cyclohexanone(I) in quasi chair conformation may also enter (almost) perfectly the cavity and occupy vertices 1-7 of $\hat{H}$,molecule II - vertices 1,2,6,7,9,10, molecule III, vertices 1-5 and 8-11. The corresponding MSD values will be $MSD_S=MSD_I=0$, $MSD_{II}=5$, $MSD_{III}=4$, but it is obvious that groups in the outer space, -OH in S, =O in I, -OCH$_3$ in II and -OCH$_2$CH$_3$ in III do not interfere with steric fit. The MSD-values "corrected" for unsuperposable atoms on steric irrelevant

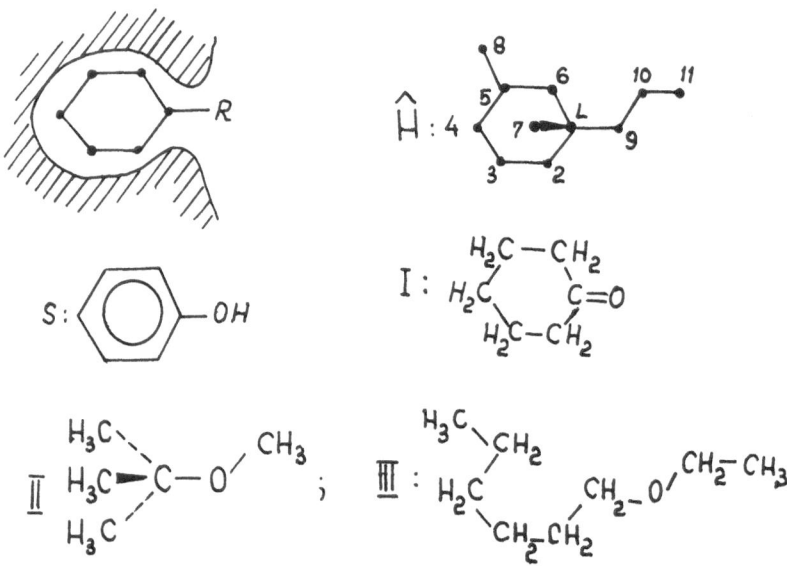

Figure 9   Receptor cavity and hypermolecule

regions, which we shall term, in what follows, MTD (Minimal Topological Differences) are $MTD_S = MTD_I = 0$, $MTD_{II} = 4$, $MTD_{III} = 2$.

The general problem is to guess, out of the A_i values for the
studied molecules, the shape of receptor cavity, receptor walls and
steric irrelevant region[20,21].

Suppose we have N molecules M_i (i=1,2,...,N) with the corresponding A_i - values. First we shall superimpose them obtaining the
topological network, the hypermolecule $\hat{H}$, with M vertices j corresponding to (approximate) nuclear positions and edges corresponding
to bonds between atoms (in at least one of the molecules). The superposition procedure is the one described at the beginning of Chap.4,
with maximal superposition on an initial standard molecule(S_o); this
should be the most active molecule of the series or an active large
molecule with complicated, but rigid steric structure. Some other
rules must also be obeyed: in enzymatic reactions, the reactive group
in each molecule M_i must occupy the same vertices j of $\hat{H}$ and also
charged groups, or hydrogen-bonding groups which chemical intuition
indicates as forming strong intermolecular forces with the receptor.
Remaining flexible groups in the N molecules should be superimposed
as much as possible one upon another; all tend to take the most fa-
vorable low energy conformation. One could also calculate nuclear co-

ordinates and contract in single vertices all nearby nuclear positions, but this was not done in the work described here.

Now, each molecule M_i will occupy certain vertices j of H, and this will be denoted by marking $x_{ij}=1$ if vertex j is occupied by molecule i, and $x_{ij}=0$ if it is not occupied. Each molecule is described by a vector $x_{i1}, x_{i2}, \ldots, x_{iM}$, or - if several low energy conformations are possible, by a set of such vectors, one for each conformation. Each vertex j may belong either to the space of the receptor cavity (marked by $\varepsilon_j = -1$), to the walls ($\varepsilon_j = +1$) or to the sterically irrelevant space ($\varepsilon_j = 0$). The set of M ternary parameters ε_j which gives the best correlation with the N values A_i represents the optimized map of the receptor, denoted by S^*, the set of vertices with $\varepsilon_j = -1$ - the "best" probable molecular shape, the optimized standard.

The corrected minimal steric difference, MTD, of molecule i can be calculated by the formula

$$MTD_i = s + \sum_{j=1}^{M} \varepsilon_j x_{ij} \qquad (1)$$

with s the number of cavity vertices (those with $\varepsilon_j = -1$). If there are molecules for which several conformations are considered, the vector $x_{i1}, \ldots, x_{iM}$ which gives the lowest MTD_i value and this lowest MTD_i is to be considered.

In the first papers[20,237,238], instead of one ternary parameter ε_j, two binary parameters δ_j, r_j were used and the formula for MTD was :

$$MTD_i = \sum_{j=1}^{M} \delta_j |r_j - x_{ij}| \qquad (1a)$$

Formulae (1) and (1a) yield the same MTD_i's if only $x_{ij}=0$ or 1 is considered, i.e., no difference between second period and higher period atoms. For $\varepsilon_j = -1$: $\delta_j = r_j = 1$, for $\varepsilon_j = 0$, $\delta_j = 0$, for $\varepsilon_j = +1$; $\delta_j = 1$, $r_j = 0$.

The optimization procedure starts from an initial set (of M) ε_j^0 values, the initial map, S^0. The regressional equation

$$A_i = \alpha - \beta\, MTD_i \qquad (2)$$

is calculated by usual regressional techniques, α possibly containing also other structural parameters (i.e., $\alpha = \alpha_0 + \alpha_1 \sigma_{1i} + \ldots$). The mean square difference of experimental and calculated activities, more exactly the expression Y, related to it :

$$y = \sum_{i=1}^{N} (A_i - \hat{A}_i)^2 \qquad (3)$$

must now be minimized also with respect to the ε_j-parameters and the receptor site is hereby mapped. This is done by a finite difference procedure : Starting from a set of parameters

$$\varepsilon_1^o, \; \varepsilon_2^o, \; \ldots, \; \varepsilon_M^o \qquad (4)$$

if ε_t^o is substituted by ε_t, Y varies with the amount $\Delta Y(\varepsilon_t)$:

$$\Delta Y(\varepsilon_t) = 2\beta(\varepsilon_t - \varepsilon_t^o) \sum_{i(x_{it}=1)} (A_i - \hat{A}_i) + \frac{\beta^2 N_{1t} N_{ot}}{N} (\varepsilon_t - \varepsilon_t^o)^2 \qquad (5)$$

The sum is performed over those molecules i in which vertex t is occupied, $x_{it}=1$. N_{1t} is the number of molecules in which -vertex t is
occupied, N_{ot} - of those in which vertex t is empty. ΔY is calculated
for all 2M monosubstitutions in the set (4) and the substitution of
ε_t for ε_t^o which yields the most negative $\Delta Y(\varepsilon_t)$ value is considered to
calculate new MTD-figures and to begin a new cycle. The procedure is
continued until all 2M variations $\Delta Y(\varepsilon_t)$ are non-negative on new mono-
-substitutions. The procedure may produce also local minima of Y in
the ε_j-space, therefore it is advisable to start from several initial
ε_j-sets(4). For the start map, one may consider the vertices occupied
by the best molecule as the cavity vertices ($\varepsilon_j = -1$) and the remaining
vertices of H either as sterically irrelevant ($\varepsilon_j=0$) or as wall vertices ($\varepsilon_j = +1$). Alternatively, one may apply first the Free-Wilson
procedure described in § 5.8 and start with a map suggested by the
corresponding regressional equation.

Certainly, this optimization method, per se, cannot say any-
thing concerning the assignment of vertices j occupied in none or in
all molecules of the series under study.

Electronic and hydrophobic-intermolecular force parameters may
be accounted within the α-term of eq.(2), or by considering intermolecular force parameters present or absent at certain vertices j of H.

The number M of ε_j-parameters to be optimized is large and
thus the reliability of such a procedure is low, even if one takes
into account that ε_j-parameters are ternary (-1, 0, +1), not continuous. The more positive are the ΔY values for the two alternative

ε_t assignments, the higher is the reliability of each ε_t assignment
in the optimized map. The ΔY-values are proportional to the increase
of square deviation between experimental and calculated activities;
this can be used in Fisher statistics to asses the reliability of the
ε_t-assignment. To obtain a realistic value for the correlative power
of the procedure, the optimized ε_j-set (4) and the regressional equation (2) must be tested on a series of compounds not used for optimization; a correlation $\hat{A}_i$ vs A_i for the test series will yield the
real predictive power. If the molecules of the test series cannot be
accommodated within the hypermolecule, one may assign $\varepsilon_j = 0$ or $\varepsilon_j = +1$
to the new vertices in a uniform manner, according to the attributions
of vicinal vertices of $\hat{H}^{21}$.

5.1 Demonstration of formula (5) for $\Delta Y(\varepsilon_t')$

According to the definition of $\Delta Y(\varepsilon_t)$

$$\Delta Y(\varepsilon_t) = Y(\varepsilon_t) - Y$$

where Y is defined by Eq.(3) and

$$Y(\varepsilon_t) = \sum_{i=1}^{N} \left[A_i - \hat{A}_i(\varepsilon_t) \right]^2 \quad \text{with} \quad \hat{A}_i(\varepsilon_t) = \alpha' - \beta MTD_i' \qquad (6)$$

MTD_i is calculated with the new set $(\varepsilon_1^0, \ldots \varepsilon_t, \ldots \varepsilon_M^0)$ but α' is considered different from α so as to conserve equality between the mean
experimental and calculated activities :

$$\sum_{i=1}^{N} \hat{A}_{i1} = \sum_{i=1}^{N} A_i \quad \text{and} \quad \sum_{i=1}^{N} \hat{A}_i(\varepsilon_t) = \sum_{i=1}^{N} A_i \qquad (7)$$

Them new MTD_i' are related to the old ones by :

$$MTD_i' = MTD_i + (S'-S) + (\varepsilon_t - \varepsilon_t^0) x_{it} \qquad (8)$$

and the new intercept, from Eq. (7) will be :

$$\alpha' = \frac{1}{N} \sum_{i=1}^{N} (A_i + \beta MTD_i') = \alpha + \beta(S'-S) + \frac{N_{1t}}{N} \beta(\varepsilon_t - \varepsilon_t^0) \qquad (9)$$

Remember that N_{it} is the number of molecules, from the total of N, in
which vertex $j=t$ is occupied. Eqs.(8) and (9), introduced into Eq.(6)
for $\hat{A}_i(\varepsilon_t)$ yield :

$$\hat{A}_i(\varepsilon_t) = \hat{A}_i - \beta(\varepsilon_t - \varepsilon_t^0)(x_{it} - N_{1t}/N)$$

which gives for $Y(\varepsilon_t)$:

$$Y(\varepsilon_t) = \sum_{i=1}^{N} [A_i - \hat{A}_i + \beta(\varepsilon_t - \varepsilon_t^0)(x_{it} - N_{1t}/N)]^2$$

In calculating the second power and with Eq.(3) for Y one obtains :

$$Y(\varepsilon_t) = Y(\varepsilon_t^0) + 2\beta(\varepsilon_t - \varepsilon_t^0) \sum_i (A_i - \hat{A}_i)^2 + 2(\varepsilon_t - \varepsilon_t^0)^2 \frac{N_{1t} N - N_{1t}^2}{N} -$$
$$(x_{it} = 1)$$
$$- 2\beta(\varepsilon_t - \varepsilon_t^0) \frac{N_{1t}}{N} \sum_{i=1}^{N} (A_i - \hat{A}_i) \ .$$

The last term is zero because of Eq.(7) and $N - N_{1t} = N_{ot}$, the number of molecules in which vertex t is unoccupied. Thus Eq.5 for $\Delta Y(\varepsilon_t)$ is finally obtained [20c].

5.2   QSAR for α-Chymotrypsine Catalyzed Hydrolyses of Esters

The specificity of α-chymotrypsine is well-known. According to Hein and Nieman[239], the substrates of the general type $R_1NH-CHR_2-COR_3$ bind to three rather hydrophobic sites of the enzyme. We choose as leading example, for MTD-mapping of receptor sites, correlations of $A_i = \log$ (bimolecular rate constant) for hydrolysis of esters with MTD and also other parameters. We choose this area, because of the wealth of experimental data, because there is also other QSAR work for comparison, and because the orientation of the molecules in the receptor site of the enzyme will be well-defined, especially for molecules $RCONH-CHR_2-COR_3$ of peptidic type, were RCONH will enter the ρ_1-site of the enzyme[240].

5.2.1   Experimental Data. Orientation of Molecules on the Hypermolecule.

Of the three sites of α-chymotrypsine, only the ρ_2-site will be mapped. We investigated[240] a series of N=14 methyl esters of N-acetylated L-aminoacids (listed by Knowles[241]), and a series of N=11 p-nitrophenol esters of carboxylic acids with rigid cycles[242] - i.e., without conformational mobility, which eliminates ambiguities in the construction of the hypermolecule and occupation of its vertices. The two sets of data are listed in Tables 26 and 27, respectively.

Table 26 Hydrolysis rates of $L-CH_3CONHCHR_2COOCH_3$

i	H_2NCHR_2COOH	A_i	$j(x_{ij}=1)$	MTD^o	MTD_{opt}
1	Trp	0.00	6,7,10-14,19,20-23	3	4
2	Tyr	-0.06	4-7,10-14,19	5	4
3	Cyclohexyl-Ala	-0.72	5-7,10-14,19	4	5
4	Phe	-1.00	5-7,10-14,19	4	5
5	$\alpha-NH_2$-Heptanoic	-1.72	11-14,19,22,23	4	5
6	Met	-2.26	10-14,19	5	7
7	$\alpha-NH_2$-Octanoic	-2.30	10-14,19,21-23	5	6
8	Leu	-2.42	7,11-14,19	5	6
9	nor-Leu	-2.53	10-14,19	5	6
10	nor-Val	-3.20	11-14,19	6	7
11	$\alpha-NH_2$-Butyric	-4.33	12-14,19	7	7
12	Ala	-5.39	13,14,19	8	8
13	Val	-5.49	12-14,19,24	8	8
14	Gly	-8.63	14,19	9	9

The hypermolecule, $\hat{H}$, (Figure 10) was constructed according to
the following considerations. The hydrolysed group $-COOCH_3$ of Table 26
and $-COOC_6H_4NO_2$ of Table 27, must occupy always the same position,
(COX in Figure 10) while the C_α-atom is always in vertex $j=14$ of $\hat{H}$.
The enzyme is a peptidase, so if there is also a peptidic bond
(CH_3CONH-group of Table 27) it should occupy always the same vertices,
in direction of site ρ_1 of chymotrypsine. Therefore, for Table 26 compounds, the NH-group of CH_3CONH occupies always vertex $j=19$, and
L-aminoacidic side chains start always towards $j=13$.

Next, the rigid cycles of the cyclocarboxylic acid esters of
Table 27 must be accommodated on $\hat{H}$. Since there exist both (condensed)
aromatic and cyclohexane rings, the last are considered in "chair"
configuration so that the planar aromatic cycles may be, approximately,
superimposed on them as well as the (equatorial) carboxylic groups
bonded to cyclohexane and to aromatic cycles. For the most active compounds of Table 27 it is natural to suppose that the decaatomic cycles
will occupy the same set of vertices ($j=5-14$), corresponding to the
best possible fit into the ρ_2-site of chymotrypsine. One should ex-

Table 27 Hydrolysis rates of $R\text{-}COOC_6H_4NO_2(p)$

i	R	A_i	$j(x_{ij}=1)$	MTD^0	MTD_{opt}
1	(D)	-0.08	3,5-14	6	6
2	(L)	-1.96	1-3,6-9,11-14	8	8
3	(S)	+1.08	5-14	5	5
4	(R)	-0.41	1-3,6-9,12-14	9	8
5	(D,L)	0.00	5-14	5	5
6		-0.59	5-14	5	5
7		-0.29	5-14	5	5
8	(D,L)	-1.44	10-19	9	8
9	(D,L)	+0.58	5-14	5	5
10		-0.10	7-9,12-14	7	6
11	$CH_3\text{-}COX$	-1.77	14	8	8

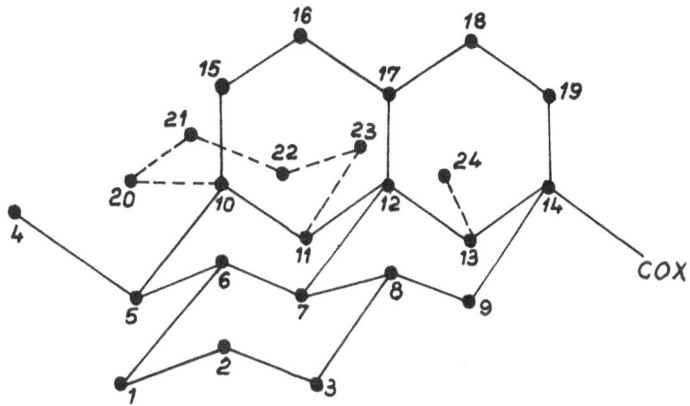

Figure 10   The hypermolecule H for molecules of
            Table 26 and Table 27.

pect the aminoacidic side chains of Table 26 compounds to be also
superimposed, as much as possible on the j=5-14 vertices.

5.2.2   Optimization Procedure. Mapping of the ρ_2 - Site

        The two series of Tables 26 and 27 will be used in two sepa-
rate optimization procedures, as X is OCH_3 and $OC_6H_4NO_2$, respectively,
for the two series and the CH_3CONH group exists only in Table 26 compounds. The two series encompass vertices (j) 4-7, 10-14, 19-24 and
1-3, 5-19, respectively, of the hypermolecule $\hat{H}$.
        The start map (standard) S_0 is the one obtained in a previous
paper[240a] by a trial and error procedure, searching something inter-
mediate between the side chains of the Trp and Tyr-derivatives, the
two most active compounds of Table 26. S_0 is depicted in Figure 11
and is characterized by the following ε_j-assignments for the vertices:

$$S_0 \begin{cases} j \ (\varepsilon = -1) \ : \quad 6,7,10\text{-}14,22,23 \\ j \ (\varepsilon = 0) \ : \qquad - \\ j \ (\varepsilon = +1) \ : \quad 1\text{-}5,8,9,15\text{-}21,24 \end{cases}$$

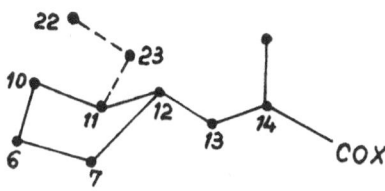

Figure 11.   Start standard, S_0.

This standard, S_0, yields
r=0.911 for the A_i-MTD^0 correlation of Table 26 and r=0.706
for the same correlation of
Table 27. The optimization
was performed by means of a
cyclic use of Equations (1,2,

3,5) on a FELIX C-256 computer[243]. For Table 26 compounds the optimized map, S (1), and the regressional equations are

$$S^*(1) \begin{cases} j(\varepsilon = -1): \ 4,6,7,10,12\text{-}15,19,22,23 \\ j(\varepsilon = 0) \ : \ 11 \\ j(\varepsilon = +1): \ 1\text{-}3,5,8,9,16\text{-}21,24 \end{cases}$$

$$\hat{A}_i = 6.412 - 1.492 \ MTD_i, \quad r=0.948, \quad N=14 \tag{10}$$

while for Table 27 compounds

$$S^*(1A) \begin{cases} j(\varepsilon = -1): \ 6,7,10,12\text{-}15,22,23 \\ j(\varepsilon = 0) \ : \ 8,11 \\ j(\varepsilon = +1): \ 1\text{-}5,9,16\text{-}21,24 \end{cases}$$

$$\hat{A}_i = 3.019 - 0.530 \ MTD_i, \quad r=0.809, \quad N=11 \tag{11}$$

The reliability of the ε_j assignments of S(1) and S(1A) is
given by the increases $\Delta Y(\varepsilon_t)$, of the mean square deviation, for the
two alternative substitutions of ε_t^0 in the optimized standard by the
ε_t-value. The vertices j occupied by all molecules or unoccupied by
all molecules of the series in Table 26 and Table 27, respectively, are
not indicated in Table 28, as the optimization procedure does not
change the initial ε_j^0-assignments for such vertices. Comparing $S^*(1)$
with $S^*(1A)$, all assignments for common vertices are identical. Combining them, the following optimized map S^* results (s_o=10):

$$S^* \begin{cases} j(\varepsilon = -1): \ 4,6,7,10,12\text{-}15,22,23 \\ j(\varepsilon = 0): \ 8,11 \\ j(\varepsilon = +1): \ 1\text{-}3,5,9,16\text{-}21,24 \end{cases} \tag{12}$$

No great changes appear with respect to the start standard S_o;
when they do ($\varepsilon_4, \varepsilon_{11}, \varepsilon_{15}$), the corresponding ΔY-values are small (0.6;
1.2 or 1.0; 0.9). Vertex j=19 should be within the cavity walls, but
if instead of ε_{15}= -1 and ε_{19}= +1 both vertices are declared sterically
irrelevant ($\varepsilon_{15}=\varepsilon_{19}=0$), the relative optimized MTD values in Table 27
remain the same.

The ρ_2-site of α-chymotrypsine should have a shape rather complementary to the start standard S_o, surrounded mostly by vertices
with ε_j= +1, i.e., corresponding to the walls of this cavity. The low-

er part of the cavity (cf. Figure 10) should also correspond to the walls (ε_j = +1); if not, the "bad", unifittable molecules would adopt other, "downwards", conformations than those of Table 26 and Table 27, to which lower MTD-values correspond, disproving the A_i-MTD_i correlation.

Table 28 Reliability of the optimization procedure.
            Results are listed as : $j(\Delta Y(-1);\Delta Y(0);\Delta Y(+1)$
            where -1, 0, +1 stand for the possible ε_j-assignments

Table 26 comp.,S^*(1)		Table 27 comp.,S^*(1A)	
4	(0.0; 0.6; 5.3)	1,2	(1.7; 0.4; 0.0)
5	(28.8; 8.1; 0.0)	3	(2.1; 0.4; 0.0)
6	(0.0; 4.7; 22.2)	5	(3.1; 0.8; 0.0)
7	(0.0; 5.8; 26.0)	6	(0.0; 0.8; 3.0)
10	(0.0; 5.5; 25.3)	7	(0.0; 0.7; 2.7)
11	(1.2; 0.0; 11.6)	8	(0.5; 0.0; 0.7)
12	(0.0; 8.2; 24.1)	9	(2.2; 0.5; 0.0)
13	(0.0; 6.9; 17.9)	10	(0.0; 1.1; 3.7)
20	(0.1; 0.1; 0.0)	11	(1.0; 0.0; 0.3)
21	(16.5; 4.4; 0.0)	12,13	(0.0; 0.8; 2.2)
22,23	(0.0; 2.6; 15.8)	15	(0.0; 0.9; 2.7)
24	(8.1; 2.0; 0.0)	16-19	(1.0; 0.1; 0.0)

5.2.3 Test of the Optimized Map S^*

The optimized map S^* will be tested on a series of 18 derivatives of D and L-aminoacids of the type

$$C_6H_5CH_2-OCONHCHR_2COOC_6H_4NO_2(p)$$

whose hydrolysis rates in α-chymotrypsine catalysis (bimolecular rate constant $\underline{v})^*$, was studied by Dupaix, Bechet, and Roucous[244]. The compounds together with MTD calculated with S^*, α-Hammett constants and π-Hansch hydrophobicities, are listed in Table 29. These 18 molecules are those of the whole series of 42, studied by[244], which present the

---

* A_i = log $\underline{v}$

hydrophobic $C_6H_5CH_2OCO$-moiety which will be fixed in the ρ_1 - site.

MTD's are calculated for the R_2-side chains. Some of the molecules of Table 29 extend beyond the hypermolecule of Figure 10, especially the D-aminoacid derivatives, in which R_2 sticks out from vertex j=14 towards j=9,8, etc. All new vertices are assigned ε_j= +1, i.e., they are considered as part of cavity walls. Electronic and hydrophobic effects are taken into account by introducing σ-Hammett and π-Hansch constants in α(i.e., $\alpha = \alpha_0 + \alpha_1\sigma_i + \alpha_2\pi_i$ in Eq.2 for A_i).

Following regressional equations, partial correlation coefficients and intercorrelation coefficients were obtained (N=18):

$$\hat{A}_i = 5.26 - 0.44 . MTD_i$$

$$r=0.777, \quad s=0.834, \quad F=11.4, \quad EV=58\% \tag{13a}$$

$$\hat{A}_i = -5.03 + 10.827\sigma_i + 0.608\pi_i$$

$$r=0.616, \quad s=0.616, \quad F=2.9, \quad EV=30\% \tag{13b}$$

$$\hat{A}_i = -0.78 - 0.41 \, MTD_i + 9.79\sigma_i + 0.58\pi_i$$

$$r=0.943, \quad s=0.441, \quad F=26.1, \quad EV=87\% \tag{13c}$$

$$r(\sigma)=0.26; \quad r(\pi)=0.41; \quad r(MTD)=0.77; \quad r(\sigma,\pi)= -0.40$$

$$r(\sigma,MTD) = -0.02; \quad r(\pi,MTD) = -0.10$$

MTD with respect to S^* explains alone more than 50% of the variance. Inclusion of σ and π allows to explain almost 90% of the variance of experimental activities. The high regressional coefficient ($\simeq$10) for the σ-Hammett constant is difficult to explain. The σ range is only 0.45-0.65. Possibly σ reflects also some polarization of the R_2-side chain which may favor binding : according to Hansch et al.[245] the α-chymotrypsine cavities are not excessively hydrophobic (MR is an even better parameter than π).

An attempt to use the MTD's in correlating the same type of results for the $RCH_2COOC_6H_4NO_2$ derivatives of Dupaix et al.[244] failed completely. R, A_i and MTD_i's are as follows :

Cl, 1.76, 8 ; I, 1.13, 8 ; CH_3O, 1.06, 7 ; $ClCH_2$, -0.04, 7 ; C_6H_5, 0.75, 9 ; H, 0.54, 9; Me-0.77, 8 ; $Cl(CH_2)_3$ 0.18, 5 ; $Cl(CH_2)_4$, 0.15, 4 ; $C_6H_5CH_2$, 1.70, 5 ; $C_6H_5(CH_2)_2$, 1.13, 6 ; $C_6H_5(CH_2)_3$, -0.15, 8 ; β-indolyl, 0.48, 8 ; β-indolyl-CH_2, 2.05,4 ; β-indolyl$(CH_2)_2$, 1.28, 7 ; $CH_3CONH(CH_2)_2$, -0.22, 4.

The possible reason is that R may enter either the ρ_1 or the ρ_2 cavity. A correlation of A_i with π and σ yields r=0.58; no sensible improvement is obtained in adding MTD.

Table 29 Correlation parameters for $C_6H_5CH_2OCONHCHR_2COOC_6H_4NO_2-p$

i	R	A_i	MTD	σ	π	$A_{i(13c)}$
1	H	0.92	10	0.57	0.00	0.69
2	L - Ala	1.18	9	0.59	0.30	1.47
3	D - Ala	0.03	11	0.59	0.30	0.65
4	L - NH_2-But	2.18	8	0.50	0.80	1.29
5	D - NH_2-But	0.25	11	0.50	0.80	0.06
6	L - Val	-0.18	9	0.45	1.10	0.57
7	D - Val	-0.97	12	0.45	1.10	-0.67
8	L - nor-Val	2.49	7	0.50	1.30	2.00
9	D - nor-Val	0.67	10	0.50	1.30	0.76
10	L - Leu	2.24	7	0.48	1.60	1.97
11	D - Leu	0.64	11	0.48	1.60	0.33
12	L - nor-Leu	2.33	6	0.50	1.80	2.70
13	L - Asn	2.40	7	0.65	-1.19	2.02
14	D - Asn	0.18	11	0.65	-1.19	0.37
15	L - Phe	3.91	6	0.57	2.43	3.75
16	D - Phe	1.68	12	0.57	2.43	1.28
17	L - Trp	3.55	5	0.57	2.44	4.16
18	L - $HysH^+$	0.83	6	0.50	-1.24	0.83

In a QSAR for all the N=42 compounds of Dupaix et al.[244] with σ, π and MSD with respect to the initial standard S_o, the value r=0.85 was obtained, while with σ, π and ES, only r=0.68. The regressional coefficient for MSD was -0.29 in this work.[240a].

5.2.4 Other QSAR on α-Chymotrypsine Catalysis

The group of Hansch[245] performed a complete QSAR-study of various kinetic constants, enzyme-substrate binding constants and inhibition constants for practically all $R_1NHCHR_2COR_3$ compounds for which such data are available. No QSAR directly related to the bimolecular, low substrate concentration rate constants was published by this group. They correlated large series of compounds, N=33-71 and even 136[246], with high m values (0.90-0.98), for large structural varieties of R_1, R_2 and R_3. Molecular refractions of the R-group, MR_1, MR_2, MR_3 turned out to be better variables than hydrophobicities, suggesting a not excessive hydrophobic character of the enzyme sites. Other parameters used were the cross product $MR_1 \cdot MR_2 \cdot MR_3$, σ-Hammett and E_S. An indicator variable with negative regressional coefficient (-0.6 to -1.5) had to be introduced for Val as H_2NCHR_2COOH, in good agreement with our optimized map S^* : the $(CH_3)_2CH$-group of Val occupies vertices j=12, 13 and 24 - the last not occupied by other aminoacids and corresponding to the cavity wall (ε_{24}= +1). The success of correlations with MR suggests a rather low rigidity of the receptor cavities; in our work regressional coefficients for MTD are also rather low (0.3-0.5) for this series.

Concerning the reliability and predictive values of the correlational equations of the Hansch group, their stability is excellent [246] : a QSAR for N=103 R_1R_2CO inhibitors of α-chymotrypsin, yields r=0.944; an extension to another 33 inhibitors, with the same structural variables, yields, for the total of N=136, r=0.940 with no significant change in regressional coefficients. Nevertheless, if the structural parameters of the series of N=103 inhibitors were used to obtain a separate correlational equation for the new N=33 inhibitors, only r=0.79 was obtained[246]. The usual technique in QSAR - to pick up the best QSAR with limited number of parameters out of thousands of tested QSAR's for the possible combinations of a rather large number of structural parameters, may also introduce some artificial correlation.

5.3 Comparative MTD and Free-Wilson Study of Anti-inflamatory Activity of Substituted Cortisol Derivatives

Relative anti-inflamatory activity of large series of cortisol derivatives are listed in the book of Shopee[247]. This series, divided into a series for optimization of N=17 compounds and a test series of N=15 compounds, is listed in Table 30 and Table 31.[248] The A_i values refer to logarithms of these anti-inflamatory activities. The tables contain also the substituents on the cortisol nucleus, the occupied vertices $j(x_{ij}=1)$ in the hypermolecule $\hat{H}$. Figure 12 illustrates the hypermolecule $\hat{H}$. The following abbreviation was employed : Acn in Tables 30 and 31 means acetonide.

The series is interesting because of its rigidity, i.e., complete lack of conformational freedom for the interesting substituents and for the fact that, due to the aliphatic character and the rather large distance between them, a good additivity of their effects is expected. The Free-Wilson model, with indicator variables for each substituent, accounts implicitly for all types (electronic, steric, hydrophobic) of additive substituent effects.

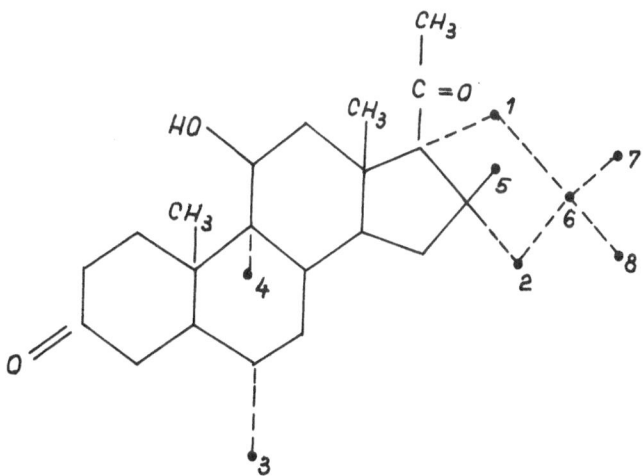

Figure 12. Hypermolecule for cortisol series with vertices j.

In the optimization procedure starting from j(ε= -1): 2,3,4,5 and j(ε=0): 1,6,7,8, the following optimized standard S^* and correlational equation are obtained for the N=17 compounds of Table 30 :

$$
S^*(4) \quad
\begin{array}{l}
j(ε= -1): \ 2,3,4,5 \\
j(ε= 0) \ : \ 1,7,8 \\
j(ε= +1): \ 6
\end{array}
$$

$$\hat{A}_i = 2.45-0.53 \ MTD; \quad r=0.756 \tag{14}$$

For the N=15 compounds of Table 31 :

$$
S^*(4A) \quad
\begin{array}{l}
j(ε= -1): \ 3,4 \\
j(ε= 0) \ : \ 1,2,5,6,7,8 \\
j(ε= +1): \ \ -
\end{array}
$$

$$\hat{A}_i = 2.32-0.83 \ MTD; \quad r=0.820 \tag{14a}$$

In the test, MTD calculated with $S^*(4A)$ for the N=17 compounds of Table 30 correlated with A_i yields r=0.673 (F=6.2, EV=42%), while MTD calculated with $S^*(4)$, for the whole series of Tables 30 and 31, of N=32 compounds, yields r=0.675 (EV=44%).

The numbering <u>k</u> for Free-Wilson variables and their correspondence with j-vertices are given in Table 32. The correlational
Free-Wilson equation for compounds of Table 32 is :

$$A_i^{FW} = 0.17 + 0.42\sigma_1 + 1.33\sigma_2 + 1.27\sigma_3 + 0.10\sigma_4 + 0.84\sigma_5 + 0.14\sigma_6 +$$

$$+ 0.81\sigma_7 + 0.53\sigma_8 + 0.02\sigma_9 - 0.80\sigma_{10} \; , \qquad r = 0.909 \qquad\qquad (15)$$

Table 30 *Cortisol derivatives and data for correlation with anti-*
-inflamatory activity. Test for optimization (Acn=acetonide)

i	Substituents	$j(x_{ij}=1)$	A_i
1	Δ_{12}	–	0.50
2	Δ_{12}, 9αF, 16α Acn	1,2,4,6,7,8	0.55
3	Δ_{12}, 6αMe	3	0.60
4	6αF, 16αAcn	1,2,3,6,7,8	0.60
5	6αF, 16αOH	1,2,3	0.70
6	Δ_{12}, 16αMe	2	0.78
7	6αCl	3	1.04
8	16αMe, 9αF	2,4	1.15
9	Δ_{12}, 16αF	2	1.20
10	Δ_{12}, 6αF, 16αAcn	1,2,3,6,7,8	1.30
11	Δ_{12}, 9αF, 16αMe	4,5	1.48
12	Δ_{12}, 6αF, 9αF, 16αOH	2,3,4	1.54
13	Δ_{12}, 9αF, 16αMe	2,4	1.60
14	Δ_{12}, 6αF, 16αMe	2,3	1.78
15	Δ_{12}, 9αF, 16αMe	2,4	1.87
16	6αF, 9αF	3,4	2.00
17	Δ_{12}, 6αCl, 9αF, 16αF	2,3,4	2.30

Table 31 *Cortisol derivatives and data for correlation with anti-*
-inflamatory activity. Test series (Acn=acetonide)

i	Substituents	$j(x_{ij}=1)$	A_i
1	–	–	0.00
2	16αMe	2	0.50
3	Δ_{12}, 9αF, 16αOH	2,4	0.50
4	6αF, 16αMe	2,3	0.70

Table 31 (continued)

i	Substituents	$j(x_{ij}=1)$	A_i
5	Δ_{12}, 6αF, 16α OH	2,3	0.70
6	9αF	4	0.90
7	9αF, 16αMe	4,5	1.15
8	6αF, 9αF, 16αOH	2,3,4	1.18
9	Δ_{12}, 9αF	4	1.30
10	Δ_{12}, 6αCl, 9αF	3,4	1.60
11	6αCl, 9αF, 16αAcn	1,2,3,4,6,7,8	1.70
12	6αF, 9αF, 16αMe	2,3,4	1.81
13	Δ_{12}, 6αF, 9αF, 16αAcn	1,2,3,4,6,7,8	2.00
14	Δ_{12}, 6αF, 9αF	3,4	2.30
15	Δ_{12}, 6αF, 9αF, 16αMe	2,3,4	2.48

The partial correlation coefficients, r_k, are also listed in Table 32.

On correlating the A_i^{FW} figures calculated for Table 31 compounds with the corresponding A_i, a factor r=0.835 (F=13.9, EV=67%) is obtained.

The test for MTD is positively significant. The test for the Free-Wilson procedure is better, but there remains a large variance (33%) which cannot be explained by an additive scheme. This may suggest large allosteric effects in the receptor site for anti-inflamatory activity[248].

Table 32 Correspondence of MTD and Free-Wilson structural parameters

j	k(FW)	ε_j	r_k
-	1(Δ_{12})	-	0.16
3	2(6αF)	- 1	0.11
3	3(6αCl)	- 1	0.29
3	4(6αMe)	- 1	-0.36
4	5(9αF)	- 1	0.40
2	6(16αMe)	-1 or 0	-0.04
2	7(16αOH)	-1 or 0	0.24
2	8(16αF)	-1 or 0	0.20
5	9(16αMe)	-1 or 0	0.33
1,2,6,7,8	10(16αAcn)	most 0	-0.35

5.4 Comparative MTD and Free-Wilson Study of Affinity for α-Adrenergic Receptors of Epinephrine Substitution Derivatives

The inhibition of ^{3}H-norepinephrine uptake into murine heart *in vivo* was studied for 30 phenyl-substituted norepinephrine derivatives by Rotman et al.[249]. The compounds are arranged into a series of N=15 compounds for the optimization procedure in Table 33, and other N=15 compounds as test series in Table 34. The corresponding hypermolecule is illustrated in Figure 13. Tables 33 and 34 contain the substituents, the k's for the Free-Wilson procedure and the occupied vertices $j(x_{ij}=1)$ of $\hat{H}$. A rotation of the phenyl ring brings j=2 into 6 and 3 into 5, the second conformation (not the used one) is also given in brackets, where it exists.

In the MTD optimization, the start standard consisted of $\varepsilon_1 = \varepsilon_3 = \varepsilon_4 = \varepsilon_5 = 0$, $\varepsilon_2 = \varepsilon_6 = \varepsilon_7 = +1$.

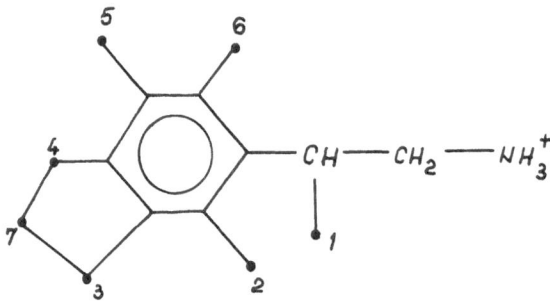

Figure 13. Hypermolecule for norepinephrine derivatives.

Table 33. Correlation of phenylethylamine derivatives with competition for α-adrenergic receptor. Series for optimization.

i	Substituents	$k(\sigma_{ik}=1)$	$j(x_{ij}=1)$	A_i
1	3-OH	3	3(5)	3.03
2	3-OH, β-OH	1,3	1,5(1,5)	2.98
3	3,5-di- OH	3,5	3,5	2.79
4	4 -OH	4	4	2.74
5	4-OH, β-OH	1,4	1,4	2.66
6	tri-2,3,5-OH	2,3,5	2,3,5(3,5,6)	2.50
7	di -2,4-OH	2,4	2,4(6,4)	2.00
8	-	-	-	2.00
9	di-3,5-OH, 4-OCH$_3$, β-CH	2,3,5,8	1,3,4,5,7	1.84

Table 33 (continued)

i	Substituents	$k(\sigma_{ik}=1)$	$j(x_{ij}=1)$	A_i
10	β - OH	1	1	1.62
11	tri-2,3,6-OH	2,3,6	2,5,6(2,3,6)	1.24
12	2-OH, β-OH	1,2	1,2(1,6)	0.44
13	tetra-2,3,4,6-OH	2,3,4,6	2,4,5,6 (2,3,4,6)	0.00
14	di-2,6-OH	2,6	2,6	0.00
15	di-4,6-OH, 4-OCH$_3$	4,6,7	3,4,6,7	0.00

Table 34 Correlation of phenylethylamine derivatives with competition for α-adrenergic receptor. Test series

i	Substituents	$k(\sigma_{ik}=1)$	$j(x_{ij}=1)$	A_i
1	di-3,4-OH	3,4	4,5(3,4)	3.08
2	tri-3,4,5-OH	3,4,5	3,4,5	2.98
3	di-3,4-OH, β-OH	1,3,4	1,4,5(1,3,4)	2.86
4	tri-2,3,4-OH	4,5,6(2,3,4)	4,5,6(2,3,4)	2.78
5	tri-3,4,6-OH	2,4,5(3,4,6)	2,4,5(3,4,6)	2.72
6	di-2,3-OH	5,6(2,3)	5,6(2,3)	2.52
7	di-3,5-OH, 4-OCH$_3$	3,5,8	3,4,5,7	2.39
8	di-3,6-OH	2,5(3,6)	2,5(3,6)	2.18
9	di-3,5-OH, β-OH	1,3,5	1,3,5	1.94
10	tetra-2,3,4,5-OH,	2,3,4,5	2,3,4,5	1.81
11	2-OH	2(6)	2(6)	1.44
12	di-3,6-OH, β-OH	1,2,5(1,3,6)	1,2,5(1,3,6)	0.51
13	tri-3,4,6-OH, β-OH	1,2,4,5 (1,3,4,6)	1,2,4,5 (1,3,4,6)	0.00
14	tri-2,4,6-OH	2,4,6	2,4,6	0.00
15	tetra-2,3,5,6-OH	2,3,5,6	2,3,5,6	0.00

Table 35   Correspondence of MTD and FW parameters

Substituents	k	j	r_k	ε_j
β - OH	1	1	0.10	0
2 - OH	2	2	-0.49	+1
3 - OH	3	3	0.26	0
4 - OH	4	4	-0.13	0
5 - OH	5	5	0.29	-1
6 - OH	6	6	-0.79	+1
3 - OCH_3	7	3,7	-0.42	0 and +1
4 - OCH_3	8	4,7	0.04	0 and +1

The optimization procedure yielded for the series of Table 33 the following optimized standard and regressional equation : [250]

$$S^* \quad \begin{array}{l} j(\varepsilon = -1) : 5 \\ j(\varepsilon = 0) : 1,3,4 \\ j(\varepsilon = +1) : 2,6,7 \end{array}$$

$$\hat{A}_i = 3.10 - 1.01 \text{ MTD} , \quad r = 0.866 \tag{16}$$

The Free-Wilson procedure for the Table 33 series yields :

$$\hat{A}_i^{FW} = 2.33 - 0.51\sigma_1 - 0.75\sigma_2 + 0.80\sigma_3 + 0.36\sigma_4 - 0.12\sigma_5 -$$

$$- 1.83\sigma_6 - 0.87\sigma_7 - 0.66\sigma_8 , \quad r = 0.927 \tag{17}$$

For the test series of Table 34, also N=15 compounds, the A_i vs. MTD calculated with S^* of Eq.(16) yields r=0.755, while the A_i vs. $\hat{A}_i^{FW}$ eq.(17) correlation yields r=0.850. The corresponding EV's are 55% for the MTD-test and 70% for the Free-Wilson test. As in the previous example, there remains 30% of the A_i variance, unexplainable by additive effects. As seen from Table 35, generally to $\varepsilon_j = -1$ there correspond positive regressional coefficients, and negative ones for $\varepsilon_j = +1$.

For another series of N=20 substitued derivatives of α-phenyl-ethylamine (with substituents at both α and β-ethylenic carbons and at the nitrogen atom) also for inhibition of noradrenaline uptake in perfused rat heart[251] , the optimized MTD procedure yields r=0.936, the Free-Wilson procedure r=0.972 (12 indicator variables). A test on N=22 compounds of the series of Rotman et al.[250] yields for A - MTD (vs. the optimized standard of the N=20 series) a value r=0.751, EV=51%, while for N=8 compounds of this test series A correlated with $\hat{A}^{FW}$ (for the N=22 series) yields r=0.792, EV=57% [250,252].

5.5 Oestrogenic Activity

Oestrogenic activity in rats, according to data listed by Malcolm, Dyson, and May[253], for a series of 20 oestrogenic steroids and stilbene derivatives, such as cis-stilboestrol, stilboestrol, oestradiol, epioestradiol, equilenine etc., was correlated with MTD and π by Popoviciu et al.[254]. The hypermolecule for this series is constructed by approximate atom per atom superposition of the 20 mole- cules. Coordinates of atoms within the sterolic cycle or the stilbene moiety were approximated, using 1.5 A for bond length and tetrahedral or trigonal bond angles. Nearby atoms, as resulted from such calcula- tions and from inspection of steric models of the molecules, were contracted into single vertices.

A hypermolecule with M=51 vertices was obtained. Hydrophobi- cities π are calculated by means of Hansch-substituent constant, start- ing from log P_{oct} values for 1,2-diphenylethane and for phenanthrene. The obtained correlational equation is :

$$\hat{A}_i = 5.54 + 1.13\pi - 0.86 \text{ MTD} ; \qquad r=0.971, \qquad s=0.39 \tag{18}$$

The optimized map, S*, hereby obtained and the correlational Equation (18) were tested on a series of 25 steroids whose oestrogenic activities in rats [255] (or the absence of such an activity, considered as A= -2) are listed in Table 37; as standard for experimental acti- vity A=2 is considered for oestrone. Only compounds not used in obtain- ing S* and Eq.(18) and with oxygen atoms bound to positions 3 and 17 of the sterolic system (or at one atom nearby - these structural cha- racteristics seem to influence very much oestrogenic activity) were considered in the test series. New atoms of the test compounds are considered in the cavity walls (ε_j = +1) or, if in the side chains bound to positions 3 and 17, sterically irrelevant (ε_j=0, as resulted for the side chains in S*). For the N=25 steroids, r=0.741, EV=53% was obtained, while for another test series of N=5 doisynolic acids, r=0.777, EV=47%.[254]

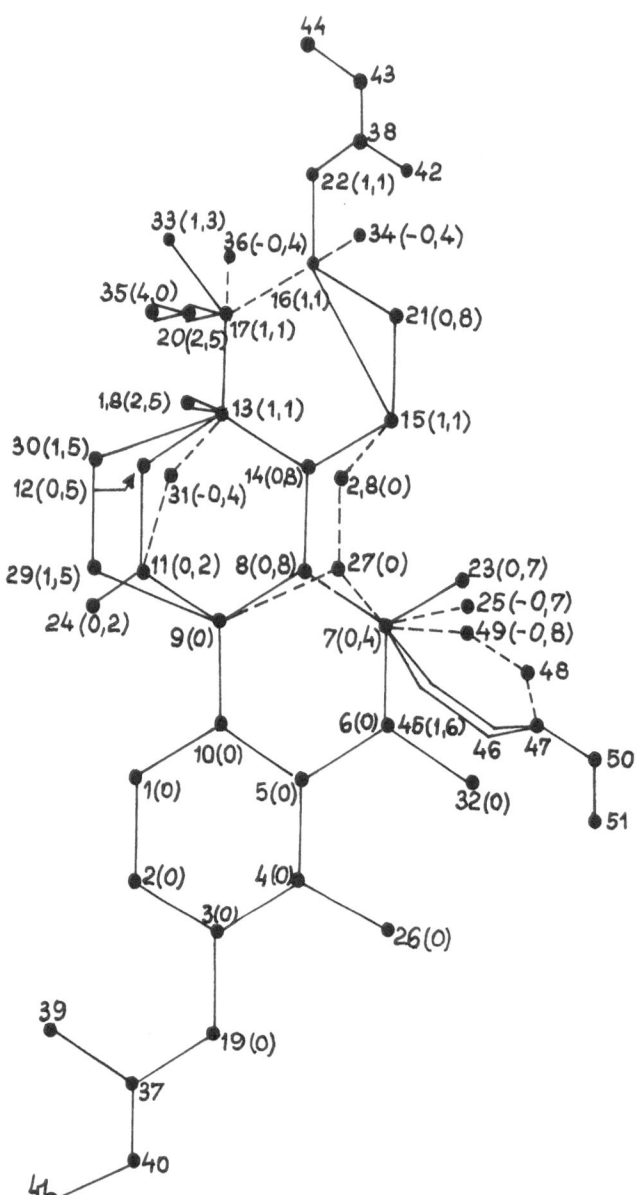

Figure 14   Hypermolecule with numbers of vertices
Figures in parentheses indicate z-coordinates in A
(heights above the paper's plane). Broken lines are below
the main plane.

The set vertices assigned in S[*] to the receptor cavity (ε_j=-1) and Eq.(18) suggest that 7β, 11α-dimethyloestradiol should be about 400 times more active than oestradiol (for intraperitoneal injection in rats).

The optimized map of the oestrogenic receptor is described by following ε_j assignments :

$$
S^* \begin{cases} j(\varepsilon = -1) : 1\text{-}7,9\text{-}12,14\text{-}20,23,24,28\text{-}31 \\ j(\varepsilon = +1) : 21,22,25\text{-}27,34,36,39,50 \\ j(\varepsilon = 0) : 8,13,32,33,35,37,38,40\text{-}49,51 \end{cases} \tag{19}
$$

The hypermolecule is illustrated in Figure 14. Values in brackets after j's indicate height (A) above the main plane of vertices. Table 36 lists the series of 20 compounds used to obtain Eq. (18). Table 37 lists the 25 test compounds which occupy also new vertices, denoted by "a" if above the steroidic system, by "b" if below, by "i" if attached to side chains at C_3 and C_{17} and by "ℓ" if attached to the steroidic systems. All new vertices were considered wall vertices (ε_j= +1), except those attached to side chains ("i") which were considered sterically irrelevant (ε_j= 0).

Table 36 Oestrogenic activity and structural parameters of steroids and stilbene derivatives.

i	Compound	A_i	π_i	MTD_i	$j(x_{ij}=1)$
1.	1,2-Di(4-hydroxyphenyl)--ethane	-3.00	3.78	14	1-5,8-10,13-17, 19-21,22
2.	Trans-1,2-di(4-hydroxyphenyl)--ethylene	-2.00	3.48	14	1-5,8-10,13-17, 19,21,22
3.	1,2-Di-(4-hydroxyphenyl)-1,2-diethyl-ethane; meso-Hexoestrol	2.90	5.38	10	1-5,7-11,13-17, 19,21-24
4.	1,2-Di-(4-propionoxy, 3-methyl-phenyl)1,2-diethyl-ethane; Mepran	3.00	6.96	12	1-5,7-11,13-17, 21-24,26,33, 37-44
5.	Cis-1,2-di-(4-hydroxyphenyl)-1,2diethyl-ethylene; ψ-Stilboestrol	1.85	5.08	11	1-5,7-11,14,15, 19,24,45-49
6.	Trans-1,2-di-(4 hydroxy-phenyl)-1,2-diethyl-ethylene; Stilboestrol	3.00	5.08	10	1-5,7-11,13-17, 19,21-24

Table 36 (continued)

i	Compound	A_i	π_i	MTD_i	$j(x_{ij}=1)$
7.	Trans-1(4-methoxyphenyl)-2(hydroxyphenyl)-1,2-diethyl-ethylene; Mestilbol	3.00	5.77	10	1-5,7-11,13-17, 19,21-24,37
8.	Trans-1,2di(4-methoxyphenyl)-1-(p-anisyl)-2-chloro-ethylene; Chlorotrianisin	3.00	7.60	13	1-5,7-10,11, 13-17,19,21,22, 37,38,45-51
9.	1,2-Di(4-hydroxyphenyl)-1,2--ethylidene-ethane; Dienoestrol	3.11	4.78	10	1-5,7-11,13-17, 19,21-24
10.	17β, 16α-Oestriol; Oestriol	1.00	1.43	7	1-20,34
11.	17β-Oestradiol; Oestradiol	3.08	2.59	6	1-20
12.	Oestrone	2.00	2.54	7	1-19,33
13.	7-Dehydrooestrone (Δ^7); Equilin	1.90	2.29	8	1-7,9,10,13-19, 27,29,30,33
14.	17-β-Dihydroequilenin($\Delta^{6,8}$)	1.30	1.99	7	1-7,9-13,15-20, 27,28
15.	Equilenin ($\Delta^{6,8}$)	0.70	1.95	8	1-7,9-13,15-19, 27,28,33
16.	Epioestradiol(17αOH)	1.60	2.59	8	1-19,36
17.	Isoequilenin (Δ^6)	1.55	1.95	7	1-11,13,15-19, 28,31,33
18.	6-Oxo-oestradiol	2.30	1.30	6	1-20,32
19.	7-Oxo-oestradiol	1.30	1.38	7	1-20,25
20.	17-Ethynyl-epioestradiol; ethynyl-oestradiol	3.11	3.01	7	1-20,35,36

Table 37 Experimental and calculated activities of some stroids (test series)

i	Compound	A_i	π_i	MTD_i	$j(x_{ij}=1)$	A_i
1.	Oestrone, cloroformiate	3.00	2.50	8	1-9,33,37,39, 40	1.49
2.	17-Methyl-oestradiol	2.30	3.09	7	1-20,36	3.02
3.	3β,17β-Dihydroxy-$\Delta^{6,8(9),10(5)}$-oestratriene	-0.30	2.05	11	1,4-7,9-13, 15-20,27, 28,a,a	-1.58
4.	17-Dihydroequilenin	0.82	2.00	7	1-7,9,10,13-20, 27,29,30	1.79
5.	(+),8,9-Dehydrooestrone	0.30	2.24	8	1-7,9-13,15-19, 27,28,33	1.20

Table 37 (continued)

i	Compound	A_i	π_i	MTD_i	$j(x_{ij}=1)$	A_i
6.	4-Hydroxy,17-equilenone	-2.00	1.95	11	1-7,9-13, 15-18,26,27,33	-1.69
7.	3-Hydroxy-16-equilenone	0.67	1.95	9	1-7,9-13, 15-19	0.02
8.	14-Dihydro,3-Hydroxy-16- -equilenone	0.45	1.65	9	1-7,9-13,15-19, 22,27,28	0.32
9.	6,7-Dihydro,3-Hydroxy- 16-equilenone	0.67	2.24	9	1-7,9-13,15-19, 22,27,28	0.33
10.	16-Methyl-equilenin	-2.00	2.25	9	1-7,9-13,15-19, 22,27,28,33	0.33
11.	D-Homooestrone	0.54	2.91	8	1-19,21,33	1.96
12.	D-Homooestradiol	0.45	2.32	7	1-20,21	2.17
13.	D-Homoequilenin	0.45	2.32	9	1-7,9-19,21, 27,33	0.44
14.	Oestrone, butyrate	2.00	2.98	8	1-19,33,37, 39-41,i	2.04
15.	Oestrone, caproate	1.67	3.98	8	1-9,33,37, 39,41,i,i,i	3.16
16.	Oestradiol, benzoate	1.89	4.04	7	1-20,37,39-41, i,i,i,i	4.09
17.	17β,16β-Oestriol	0.85	1.43	7	1-20,a	1.15
18.	Octahydrooestrone	-2.00	3.40	12	1-11,13,17,19, 28,31,b,b,b,b	-0.92
19.	2-Hydroxy-oestrone	-2.00	1.87	8	1-19,33,ℓ	0.79
20.	19-Homoequilenin	0.37	2.45	9	1-7,9-13,15-19, 27,28,33,a	0.58
21.	19-Bis-homoequilenin	0.15	2.95	10	1-7,9-13,15-19, 27,28,33,a,a	0.29
22.	19-Tris-homoequilenin	0.75	3.45	11	1-7,9-13,15-1, 27,28,33,a,a,a	-0.01
23.	1-Methyl-oestrone	1.67	3.04	8	1-9,33,ℓ	2.10
24.	14-Isoequilenin	-2.00	1.95	15	1-7,9-11,13,19, 27,28,31,33, b,b,b,b	-4.12
25.	Cis-Oestrone	-1.15	2.54	10	1-14,17-20,b,b	-0.21

5.6. Dihydrofolate Reductase Inhibition

Dihydrofolate reductase inhibition (- lg of inhibition constants) was correlated with MTD and π in a previous paper[21] for 15 di-
aminopyrimidines substituted in position 5 with aliphatic, cycloali-
phatic (ex. adamantyl) and aromatic radicals (Table 38). The hypermolecule is depicted in Figure 15 and contains M=25 vertices.

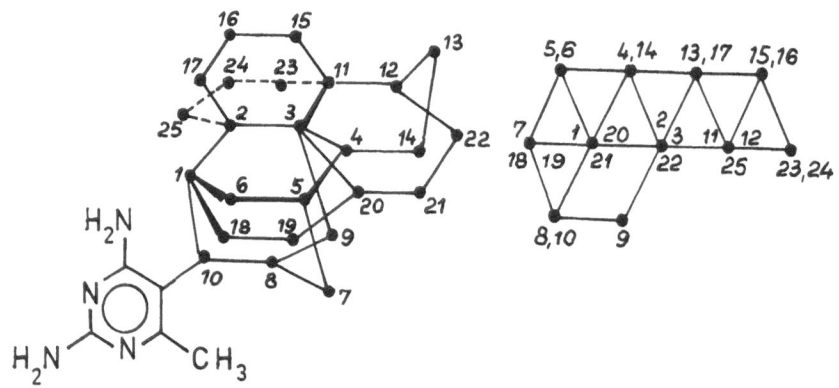

Figure 15   Hypermolecule for dihydrofolate reductase inhibition.

The optimized map and regressional equation are :

$$
S^* \begin{cases} j(= -1) : 1\text{-}6,10,11,15,21,22 \\ j(= 0) : 7\text{-}9,12\text{-}24,16\text{-}20 \\ j(= +1) : 23\text{-}25 \end{cases}
$$

$$A_i = 9.22+0.36\ _i-0.64\ MTD_i; \qquad r=0.981, \qquad s=0.29 \qquad (20)$$

$$r(MTD)=-0.95; \quad r(\pi)=0.77; \quad r(MTD,\pi)=-0.78$$

The most potent dihydrofolatereductase inhibitor as suggested
by the optimized map S^* (20) should be the 5-derivative of 2,4-diamino-
-6-methyl-pyrimidine with following substituents : 1(-axial) methyl,
3(-equatorial) ethyl, cyclohexyl (chair conformation).

The result can be tested by comparison with results of Silipo
and Hansch[256], who, for a series of N=244 derivatives of the type 4,6-
diamino-1,2-dihydro-2,2-dimethyl-1-(X-phenyl)-s-triazines, also for dihydrofolate reductase inhibition, obtained the following regressional
equation :

$$A_i = 6.49+0.68\pi_3-0.118(\pi_3)^2+0.230\ MR_4 -$$
$$-0.0243\ MR_4^2+0.238\ I_1-2.53\ I_2-1.99\ I_3 +$$
$$+0.88\ I_4+0.69\ I_5+0.70\ I_6; \quad r=0.923; \quad s=0.38 \qquad (21)$$

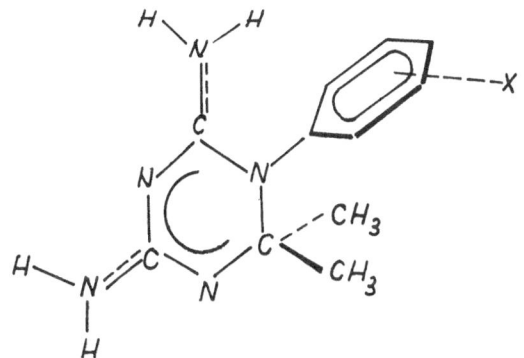

All variations are expressed
by different X-substituents
in the phenyl. π_3 and MR_4
are hydrophobicity and molar
refraction for meta and para
substituents, I_2 marks the
presence of an ortho substituent, $I_1\ldots,I_6$ are other
indicator variables. Thus,
rather small meta and para
substituents ($\pi_3 \lesssim 2.9$,

Figure 16 Compounds of Silipo
and Hansch (Eq.21).

$MR_4 \lesssim 4.7$) enhance inhibitory activity, while ortho
substituents depress it.

To compare the result of (20) with that of (21) one must take
into account the planarity, due to conjugation, of the triazine ring
and the amino groups. The phenyl ring of the Silipo and Hansch compounds should thus be perpendicular to the triazine cycle and occupy
vertices j=1,2,3,20,19,18 of the hypermolecule of Figure 15. Ortho
substituents will stick out from j=2 towards j=25, which represents a
receptor wall vertex : ε_{25}= +1 and the regressional coefficient for I_2
is -2.53. Meta substituents stick out from j=3 towards j=11, para
substituents from j=20 towards j=21; both vertices correspond to the
receptor cavity, ε_{11} = -1, ε_{21}= -1 and to positive regressional coefficients, +0.68 and +0.23 in Eq.(21). The N=244 compounds of Silipo
and Hansch cover a much wider space than that corresponding to the
hypermolecule of Figure 15, but for small ortho, meta and para substituents, the sign of regressional coefficients of Eq.(21) agrees
with the assignment of vertices in S*, Eq.(20). One should also remember that the "left hand" side (with respect to the phenyl vertices
1-3, 18-20), not depicted in Figure 15, cannot be "better" that the
"right hand"-side (vertices j=11-13, 15-17, 23-15): if it would be,
some of the N=15 compounds corresponding to Eq.20 would rotate to
"left" giving lower MTD's, a fact which is excluded by the optimization method of[21]. As another test, in ref.[21], $\hat{A}_i$'s were calculated
for three 2,6-diaminopurine derivatives substituted in position 8.
Purine ring atoms 7 and 9 were considered to occupy j=1 of $\hat{H}$ and the
position of the CH_3 group, the space exterior to $\hat{H}$ as sterically irrelevant. For the H, adamantyl and methyl-adamantyl derivatives, the
experimental figures, A_i, are 2.91, 5.32 and 4.88, MTD's 10, 8 and 8

(the last two molecules occupy also j=21,22 of $\hat{H}$), the calculated $\hat{A}_i$:
2.29; 5.07 and 5.25.

An optimization with MTD π_3 and π_3^2 on a simplified, planar
hypermolecule with M=80 vertices yielded, for the series of Silipo and
Hansch, r=0.86, but, as no test was performed, one cannot know how
much thereof is due to parameter adjustment[257].

*Table 38   Dihydrofolate reductase inhibitors (5-substituted
2,4-diamino-6-methyl-pyrimidines)*

i	R	A_i	π_i	MTD_i	$j(x_{ij}=1)$
1.	CH_3	3.00	0.56	10	1
2.	C_2H_5	3.10	1.02	9	1,2
3.	$n\text{-}C_3H_7$	5.13	1.55	8	1-3
4.	$t\text{-}C_4H_9$	5.49	1.98	7	1,2,6-10
5.	$cyclo\text{-}C_6H_{11}$	6.89	2.51	5	1-6
6.	$n\text{-}C_5H_{11}$	6.15	2.55	6	1-5
7.	$n\text{-}C_6H_{13}$	6.32	3.05	6	1-5,7
8.	Adamantyl	8.22	3.30	4	1-10
9.	$n\text{-}C_7H_{15}$	6.66	3.55	6	1-5,7,8
10.	$n\text{-}C_8H_{17}$	6.72	4.05	6	1,3,5-9,11
11.	$n\text{-}C_{10}H_{21}$	7.22	4.55	5	1,3,5-9,11-13
12.	$CH_2\text{-}cyclo\text{-}C_6H_{11}$	6.38	2.81	6	1-3,11,15-17
13.	$(CH_2)_2\text{-}cyclo\text{-}C_6H_{11}$	6.57	3.31	6	1-4,11-14
14.	β-Naphthyl	7.15	3.37	5	1-3,11,12,18-22
15.	α-Naphthyl	4.25	3.37	10	1-3,11,18-20,23-25

For the $n\text{-}C_{10}H_{21}$ derivative (i=11) the conformation indicated in
Table 38 was used in the optimization procedure; this conformation
yields MTD=6. The value r=0.981 was obtained with MTD=5 for this derivative, which corresponds to the "best" low energy conformation
$j(x_{ij}=1)$:1,3,5-9,11,15,16.

5.7   Other Correlational Work on MTD

Haptenes.   The affinity for an antibody against 3-azopyridine
of N=49 substitution derivatives of pyridine and succinylamidobenzene
was correlated with MTD and some other structural parameters by
Badilescu and Simon[238]. Hydrophobicity π, molecular weight of sub-

stituents MW and presence or absence (n=1 or 0) of a proton-accepting
lone electron pair in the position of the pyridinic N atom were also
considered. In a simple correlation with MSD and n, r=0.710 was obtained; from the tested intermolecular force constants only n is highly significant. The MTD procedure improved the correlation to r=0.883.
Hansch and Moser[258] performed a QSAR work on the same group of haptenes,
but separately for pyridine-substitution derivatives (N=22; r=0.92;
k=4) and for benzene substitution derivatives (N=20; r=0.95; k=5).They
used molecular refraction, MR, hydrophobicity, σ-Hammett constants separately for ortho, meta, and para substituents, as structural variables, but only as a total of k=4 or 5 (k is here the number of
structural parameters used in the regressional equation). Their regressional coefficients agree with the ε_j-assignments in the hypermolecule
of [238], for pyridine derivatives but not for benzene derivatives.
Both papers agree in that the antibody receptor cavity should be
rather polar.

   Phenyl-methyl-carbamates, acetylcholinesterase inhibitors.
Acetylcoline-esterase inhibition was correlated with phenyl substituent parameters (MR, π, HB, EC and MTD) for N=97 derivatives. Correlation with intermolecular force parameters yields r=0.764, addition
of MSD rises correlation to r=0.800[227] and the use of the MTD optimization procedure, to r=0.876 (r(MTD)= -0.51)[259]. Test trials of the
optimized receptor map and correlational equation were rather unsa-
tisfactory; significant $\hat{A}_i$ - A_i correlations could be obtained only
for a series of N=8 phenylthiophosphates (r=0.79) and of N=8 hetero-
cycle-methyl-carbamates (r=0.52).

5.8 Vertices as Indicator Variables

   The hypermolecule can be used as basis of a Free-Wilson type
procedure in which each vertex corresponds to an indicator variable
(1 if occupied, 0 if empty in M_i) and a correlational equation is set
up by the usual regressional technique[260]. Groups of vertices uniformly occupied or unoccupied, in the studied set of molecules, will
appear as single indicator variables.
   As an example we present the parallel Free-Wilson and MTD
treatment of a set of N=13 carboxylic acids, inhibitors of α-carboxy-
peptidase[237]. The molecules, their inhibitory activity (A_i = -log K_I,
K_I-inhibition constant, mol/ℓ) as given by Polglase, Lumry, and
Smith[261] are listed in Table 39. The hypermolecule is depicted in

Figure 17, the occupied vertices $j(x_{ij}=1)$ and Free-Wilson variables $k(\sigma_{ik}=1)$ are indicated also in Table 39, the correspondence of j and k-parameters in Table 40.

Table 39 Correlational data for carboxypeptidase inhibitors

i	M_i	A_i	$j(x_{ij}=1)$	$k(\sigma_{ik}=1)$
1	C_2H_5COOH	1.00	1,2	-
2	$n\text{-}C_3H_7COOH$	2.30	1,2,6	3
3	$n\text{-}C_4H_9COOH$	2.57	1,2,5,6	2,3
4	$n\text{-}C_5H_{11}COOH$	2.20	1,2,5,6,10	2,3,7
5	$(CH_3)_2CHCH_2CH_2COOH$	2.57	1,2,5-7	2-4
6	C_6H_5COOH	0.83	1,2,6,7,11,12	3,4,8
7	$C_6H_5CH_2COOH$	2.34	1-6,13	1-3,9
8	$C_6H_5CH_2CH_2COOH$	2.92	1,2,5-10	2-7
9	$C_6H_5(CH_2)_3COOH$	1.70	1,2,5,6,10,13-16	2,3,7,9-11
10	$\beta\text{-}C_8H_6N\ CH_2COOH$	3.11	1-10	1-7
11	$\beta\text{-}C_8H_6N(CH_2)_2COOH$	2.26	1,2,5-8,10,13-16	2-5,7,9-11
12	$\beta\text{-}C_8H_6N(CH_2)_3COOH$	1.48	1,2,5,6,9,10,13,14,16-19	2,3,6,7,9,10
13	$\beta\text{-}C_{10}H_7CH_2COOH$	1.66	1,2-6,10,13-19	1-3,7,9-11

$\beta\text{-}C_8H_6N$: β-indolyl ; $\beta\text{-}C_{10}H_7$: β-naphthyl

Vertices j=1 and 2, occupied in all molecules are not considered as indicator variables, vertices j=11 and 12 occupied only in molecule i=6, correspond to k=8, vertices j=14,16, occupied in i=9,11-13 to k=10, vertices j=17-19, occupied in i=12,13, to k=12.

Table 40 Vertex - "Free-Wilson" parameter correspondence.
The second column group of values represent $\varepsilon_j(\Delta Y(-1),\Delta Y(0),\ \Delta Y(+1))$

j	$\varepsilon_j(\Delta Y(-1),\Delta Y(0),\Delta Y(+1))$	k	α_k	$r(\sigma_k)$
1,2	-1(0.0, 0.0, 0,0)	-	-	-
3,4	0(0.7, 0.0, 0.6)	1	+0.075	0.244
5	-1(0.0, 1.0, 3.4)	2	+0.270	0.570
6	-1(0.0, 0.8, 2.1)	3	+1.300	0.463
7	0(0.9, 0.0, 0.8)	4	+0.000	0.314
8	-1(0.0, 0.7, 2.7)	5	+0.617	0.565

Table 40 (continued)

j	$\varepsilon_j(\Delta Y(-1),\Delta Y(0),\Delta Y(+1))$	k	α_k	$r(\sigma_k)$
9	-1(0.0, 0.5, 2.2)	6	+1.600	0.353
10	+1(2.4, 0.3, 0.0)	7	-0.370	0.190
11,12	+1(1.3, 0.4, 0.0)	8	-1.470	-0.536
13	0(0.5, 0.0, 1.2)	9	-0.305	-0.218
14	+1(1.9, 0.2, 0.0)	10	+0.050	-0.296
16	-1(0.0, 1.4, 4.2)			
15	+1(1.3, 0.0, 0.0)	11	-0.302	-0.163
17	0(0.3, 0.0, 0.2)	12	-0.625	-0.256
18,19	+1(1.1, 0.3, 0.0)			

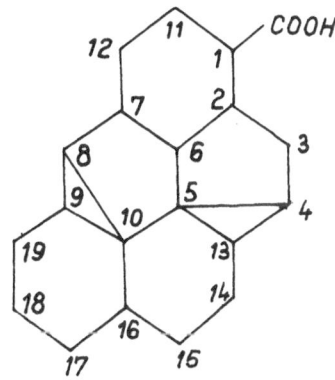

Figure 17 Hypermolecule for carboxypeptidase inhibitors.

In the MTD-procedure, the optimized map S^* and the correlational equation are :

$$S^* \begin{cases} j(\varepsilon=-1) : 1,2,5,6,8,9,16 \\ j(\varepsilon= 0) : 3,4,7,13,17 \\ j(\varepsilon= +1): 10,12,14,15,18,19 \end{cases} \quad (22)$$

$$\hat{A}_i = 4.139 - 0.527\ MTD_i ; \quad r=0.950$$

while the corresponding Free - Wilson equation is :

$$\hat{A}_i^{FW} = 1.000+0.075\sigma_1+0.270\sigma_2+1.300\sigma_3+0.000\sigma_4+0.617\sigma_5+0.160\sigma_6 -$$
$$-0.370\sigma_7-1.470\sigma_8-0.305\sigma_S+0.050\sigma_{10}-0.302\sigma_{11}-0.625\sigma_{12} ; \quad r=0.999$$
$$(23)$$

Table 40 lists also the $\Delta Y(\varepsilon_t)$'s for alternative ε_j-assignments as compared to the optimized map S^* (22) and partial correlation coefficients, $r(\sigma_k)$, for the indicator variables. There is a rather good correspondence between ε_j-assignments and the sign of the corresponding regressional coefficients α_k of Eq.23 and Table 40.

The Free-Wilson equation reflects, in principle, also the ridigity of the receptor in the vicinity of the corresponding vertex. It can also suggest a start-map for the MTD-procedure: for highly positive α_k's, there correspond $\varepsilon_j = -1$, for highly negative α_k's,

ε_j= +1 and for small (positive or negative) or zero α_k's, ε_j=0. Nevertheless, due to the large number of continuous adjustable parameters, the reliability of the Free-Wilson procedure is very low. It would also be difficult to take into account several conformations for one molecule in this procedure.

6. MCD - MONTE CARLO VERSION FOR MINIMAL STERIC DIFFERENCE

The previous two chapters described the MSD and MTD techniques for accounting steric effects in receptor-effector interactions. The correlational power of these techniques was illustrated by several examples of QSAR. An analysis of these techniques emphasizes the following features :

(i) MTD is a versatile topological method which can be used for series of chemical compounds of largely differing structures. Nevertheless the MTD-techniques, as well as the simple MSD version, becomes ambiguous in some situations, as for example series containing polycyclic molecules or cycles of different sizes : the superposition procedure cannot be exactly defined.

(ii) MSD and also MTD use the concept of standard molecule, considered as a complementary copy of the receptor cavity and that of a hypermolecule (implicitly for MSD) which fixes in a convenient way the "topological coordinates" of the molecules in the correlated series. The "best" molecule is used as standard in MSD and the optimization procedure used in MTD starts from it. The use of the term "standard" in this acception may be criticized from the phenomenologic viewpoint. Its use and that of the hypermolecule may be somewhat restrictive in obtaining regressional equations but they are required by the topologic essence of MSD and MTD.

(iii) In order to superpose the molecule on the standard, one uses the rule of obtaining minimal nonsuperposable volumes. This rule is not always a sufficient condition, as the superposition "translates" for the MTD formalism also the problem of conformation attacking the receptor. This problem is complex (correct position for bonds to be cleaved, juxtaposition of certain intermolecular bond-forming groups) and requires supplementary informations.

The Monte Carlo version of minimal steric difference, designed hereafter as MCD, improves the computation of nonoverlapping volumes in the standard-molecule superposition by giving up the topological formalism for the sake of the metric one[17-19]. The ambiguities emphasized under (i) are hereby eliminated. As far as it accepts the other assumptions[14] of MSD- and MTD-techniques, MCD will present the same inconveniences, mentioned under items (ii) and (iii), as the MTD-technique.

6.1 The Method to Calculate Non-overlapping Volumes

The mathematical method used in the MCD-technique is the Monte Carlo method[261-264]. Random numbers $\xi \in (0,1)$ with random distribution within this interval are used from which the numbers $\xi \in (a,b)$, $a,b \in R$ are obtained following relation (1)

$$\xi' = a + (b - a)\xi \tag{1}$$

Numbers ξ' are randomly distributed within the (a,b) interval.

The precision of Monte Carlo calculations is conditioned by the quality of the random number generator and by the length of the random number sequence used in calculations. Random number generators existing in libraries of mathematical computer systems are satisfying high precision requirements. Thus, the autocorrelation test (with χ^2 statistics) for generators RANDOM (IBM) and ALEAT (CII) produced results summed up in Table 41.

In order to determine the length L of the sequence of random numbers ξ' which assures calculation of MCD figures with a required error ε, one may adopt a heuristic method - for example an expensive self-consistent type method, or a function

$$\varepsilon = f(\alpha, L) \tag{2}$$

can be calibrated, which implements itself into the program, and the length L_o of the sequence which assures the required precision ε_o is determined automatically. The significance of α will be explained below.

Table 41 RANDOM and ALEAT performances.
         Calculations were performed with an IRIS-50 computer

Generator	Autocalculation Dimension of space	χ^2 (for 64 freedom degrees)	Time for generation of 10^6 numbers
ALEAT	1	75.03	
	2	81.63	358 sec.
	3	63.08	

Table 41 (continued)

Generator	Autocalculation		Time for generation of 10^6 numbers
	Dimension of space	χ^2(for 64 freedom degrees)	
RANDOM	1	88.27	
	2	61.68	364 sec.
	3	69.36	

6.1.1 Description of Molecules

In order to calculate MCD, the molecules are described by the cartesian coordinates and van der Waals radii of their atoms.

For calculations of coordinates the standard values of geometric parameters can be used (bond lengths and bond angles) and van der Waals radii listed in ref[265] or [266]. The way one achieves the superposition is implicitly specified by the atomic coordinates: all molecules of the series are represented in the same cartesian coordinate system.

Let us consider molecules M_α and M_β out of a given series to be used for QSAR, one of them being the standard. M_α consists of N_α atoms of coordinates ($A_{\alpha I}$, $B_{\alpha I}$, $C_{\alpha I}$) and van der Waals radii $R_{\alpha I}(I=1,2,...,N_\alpha)$. M_β consists of N_β atoms of coordinates ($A_{\beta j}$, $B_{\beta j}$, $C_{\beta j}$) and van der Waals radii $R_{\beta j}(j=1,2,.....,N_\beta)$. The van der Waals envelopes of M_α and M_β are thus described by the collection (3) of spheres :

$$(\xi'_x - A_{\alpha I})^2 + (\xi'_y - B_{\alpha I})^2 + (\xi'_z - C_{\alpha I})^2 < R^2_{\alpha I} \qquad (3)$$

$$I = 1,2,..., N$$

and, respectively, by collection (4) of spheres :

$$(\xi'_x - A_{\beta J})^2 + (\xi'_y - B_{\beta J})^2 + (\xi'_z - C_{\beta J})^2 < R^2_{\beta J} \qquad (4)$$

$$j = 1,2,..., N$$

i.e., the point P_ξ, of coordinates (ξ'_x, ξ'_y, ξ'_z) is within the van der Waals envelope of M_α if it satisfies at least one of Eqs.(3), and within that of M_β if it satisfies at least one of Eqs.(4).

6.1.2 Calculation of MCD

Following its definition, MCD is the total volume of the un-superposable van der Waals envelopes which do not overlap at the standard molecule superposition.

Point $P_{\xi'}$, $(\xi'_X, \xi'_Y, \xi'_Z) \in R^3$ belongs to a nonoverlapping zone of van der Waals envelopes of molecule M_α or M_β if it satisfies a single inequality of system (5) with $N_\alpha + N_\beta$ inequalities :

$$(\xi'_x - A_{\alpha I})^2 + (\xi'_y - B_{\alpha I})^2 + (\xi'_z - C_{\alpha I})^2 < R^2_{\alpha I}$$

$$(\xi'_x - A_{\beta I})^2 + (\xi'_y - B_{\beta J})^2 + (\xi'_z - C_{\beta J})^2 < R^2_{\beta J}$$

(5)

ξ'_X, ξ'_Y, ξ'_Z , the coordinates of the random point $P_{\xi'}$ are generated with a random number generator.

If point $P_{\xi'}$ satisfies at least one type "α" and one type "β" equation of system (5) it falls into a zone in which the van der Waals envelopes of the two molecules overlap. If $P_{\xi'}$ satisfies neither of the equations of system (5) it falls outside the envelopes of both molecules.

The manifold of points $P_{\xi'}$, satisfying a single Equation (5) is denoted by N_I, the manifold of points satisfying at least one "α" and one "β"-equation or neither of Equations (5), by N_A. The manifold N_T is $N_I \cup N_A = N_T$.

Let α be the volume of the parallelipiped which contains the superposed van der Waals envelopes of molecules M_α and M_β. Following the methodology of the Monte Carlo technique, MCD is calculated by relation (6) :

$$MCD = \alpha \frac{|N_I|}{|N_T|}$$

(6)

where $|N_I|$ denotes the cardinal of manifold N_I and, respectively, N_T. Figure 18 illustrates the principle of MSD-calculus. MCD is to be used as steric parameter in correlational equations in the same sense as E_S, MSD or MTD.

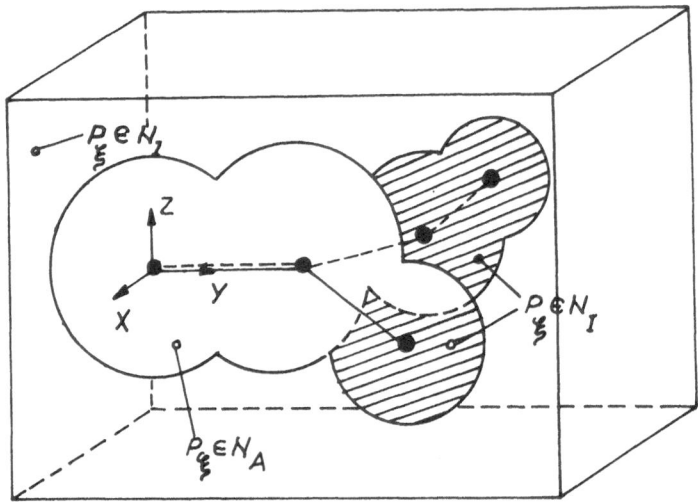

Figure 18   Calculation of MCD. The skeleton of molecule M_α is marked by (——), that of M_β by (---). The volume of the darkened zone represents MCD.

6.1.3   Computer Implemented MCD-Calculations. The MCD - Program

The calculation method described in the preceeding paragraph is implemented by the MCD-program which has three main sections.

1) Calculation of cartesian atomic coordinates. For this purpose a QCPE-program[267] was adapted to our scope and our calculation system. This section lists the cartesian atomic coordinates of the molecules under consideration.

2) Calculation of MCD with formula (6), selecting as standard the most active molecule of the series. This section lists the calculated MCD values.

3) Multiparametric regressional analysis. Equations (7) and (8) are derived with the corresponding statistics :

$$\hat{A} = \alpha - \beta \, MCD \tag{7}$$

$$\hat{A} = \alpha - \beta \, MCD + \sum_{j=1}^{k} c_j \sigma_j \tag{8}$$

$\hat{A}$ is the calculated activity, σ_j-other structural parameters. MCD requires $k \geq 1$.

A detailed presentation of MCD - block scheme functionality and interaction of subroutines structure of entrance data set, complete listing of the program is given in Appendix 9.2 and thus this program can be used without difficulties.

6.1.4 Applications

One illustrative example of MCD will be given here. Further work will be published elsewhere. The example refers to substituent effects in haptenes, in their interaction with the anti-p,p'-azo-phenyl-azo)benzoate antibody[268,269]. Experimental values are taken from ref.[279]. Table 42 lists relative binding constants and MCD values with p-chlorobenzoate, the derivative with highest affinity, as standard. The log K_{rel} correlation gives r=0.68, which for the N=11 compounds has a statistical significance above 95% [271].

In another study[19] we correlated enzymatic hydrolysis rates, v, of N-glycosidic bonds in nucleosides and of Michaelis constants K_M with MCD and simple MSD. The log v_r vs. MCD correlation yields r=0.434 for N=10 compounds (r=0.572 for log v vs. MSD), the log K_M vs MCD, r=0.884 for N=7 compounds (r=0.899 for log K_M vs MSD). The lower r's for the correlations with MCD are probably due to the selected orientation of molecule vs. standard in MCD calculations in ref.[19], which does not allow maximal superposition.

Table *Steric effects in affinity for anti-p, p'-azophenylazo)benzoate*

Nr.	Substituent	K_{rel}	MCD(A^3)
1.	3-Cl	0.43	16.2
2.	4-Cl	5.30	0.0
3.	2-NO$_2$	0.01	13.7
4.	3-NO$_2$	0.12	16.4
5.	4-NO$_2$	1.80	1.4
6,	3-NO$_2$, 4-Cl	0.41	19.4
7.	2-NO$_2$, 5-Cl	0.07	34.6
8.	3,5-di-NO$_2$	0.024	38.3
9.	2,4-di-NO$_2$	0.014	13.7
10.	2,5-di-NO$_2$	0.013	31.4
11.	2,3,6-tri-NO$_2$	0.01	20.4

7. METRICES IN BIOCHEMISTRY. THE METRIC INDUCED BY MINIMAL STERIC DIFFERENCES

Consider a set of objects $P = \{p,q,r,\dots\}$. Function ρ is defined according to relation (1)[272,273]

$$\rho : \quad P \times P \rightarrow R_+ \qquad\qquad (1)$$

where "$\times$" denotes a cartesian product and R_+ is the real positive semi-axis. One calls[279] ρ as dissimilarity coefficient if it satisfies conditions (2) :

$$\rho(p,q) \geq 0, \quad \text{for all } p,q \in P \qquad\qquad (2a)$$
$$\rho(p,p) = 0, \quad \text{for all } p \in P \qquad\qquad (2b)$$
$$\rho(p,q) = \rho(q,p), \quad \text{for all } p,q \in P \qquad\qquad (2c)$$

If ρ satisfies also the property (3) of the triangle :

$$\rho(p,r) \leq \rho(p,q) + \rho(q,r), \quad \text{for all } p,q,r \in P \qquad\qquad (3)$$

then ρ is a metric on the set P.

The usefullness of defining a metric or a dissimilarity coefficient, ρ, on the pertinent set P is evident; by ρ one can perform a quantitative comparison of the elements $p \in P$. $\rho(p,q)$ is the measure for dissimilarity between the (mathematical) objects p and q (of the differences in the sense of the definition of ρ).

The importance of ρ in the QSAR-context is revealed by relation (1); it gives numeric values to the independent variables (parameters) of the correlational equation. Interesting results were obtained in connection with distances of two protein sequences (Beyer et al.[275-277], Ulam[278] and others[279,280], Fitsch[281,282]).

In what follows we are interested in the manifold M :

$$M = \{m_i \mid m_i = \text{molecules }\}_{i \in ICN} \qquad\qquad (4)$$

termed as fundamental space, m_i being the points of the space and N the manifold of natural numbers. Taking into account the simplifications in the original version[14] of MSD, the following assertions hold :

Definition 1. The points of a cubic lattice L in the euclidian space E_3 represent the topological coordinates for any $m_i \in$ M (each atom in m_i corresponds to a point of lattice L).

Lemma 1. For any $m_i \in$ M there exists a sublattice $l_i \in$ M which describes the chemical structure of molecule m_i.

Lemma 2. M and P(L) are isomorphic : M $\cong$ P(L). Actually M and P(L) are numerable.

P(L) is the manifold of all sublattices of lattice L. Demonstrations of these two lemmas are obvious.

Corollary. Operating with $m_i \in$ M is equivalent to operating with $l_i \in$ L. To the sublattice vertices $l_i \in$ L the value W_{rpq} is attached, the approximate van der Waals volume of the atom corresponding to vertex (rpq) of l_i. One can introduce :

Definition 2 : The compositional law $\odot$

$$\odot : P(L) \times P(L) \rightarrow R_+ \tag{5}$$

$\odot$ is given by relation (6)

$$l_i \odot l_j = \sum |W_{rpq}^i - W_{rpq}^i| \tag{6}$$
$$\text{all } r,p,q$$

$l_i, l_j \quad P(L), \quad r= 1 \div R, \quad p= 1 \div P, \quad q= 1 \div Q, \quad R,P,Q \in N$

Observations : (6) is always possible because l_i and l_j can always have the same dimensions, for example RXPXQ, by filling the corresponding sublattice with zero's.

$\odot$ is commutative : $\quad l_i \odot l_j = l_j \odot l_i \tag{7}$

$\odot$ is nonassociative : $l_i \odot (l_j \odot l_k), (l_i \odot l_j) \odot l_k$ is senseless because $l_i, l_j, l_k \in$ L, while $(l_i \odot l_j), (l_j \odot l_k) \in R_+$. Thus relation (8) is valid

$$MSD(m_i,m_j) = l_i \odot l_j = l_j \odot l_i = MSD(m_j,m_i) \tag{8}$$
$$m_i,m_j \in M \text{ and } l_i,l_j \in P(L)$$

Proposition : MSD induces a Hausdorff space on M.
Really, if function ρ is defined as :

$$\rho : P(L)XP(L) \rightarrow R_+$$

one has :

$$\rho(1_i,1_j) = \sqrt{(1_i \ominus 1_j)^2} \tag{9}$$

It is easily observed that ρ, as defined by (9), verifies conditions (2). Thus ρ constitues a dissimilarity coefficient, and as ρ verifies also relation (3), ρ is a metric and $(P(L),\rho)$ a metric space. Following a topological theorem[283], $(P(L),\rho)$ is a Hausdorff space. From $M \sim P(L)$ it results that (M,ρ) is also a Hausdorff space, the theorem being demonstrated.

Observations 1) In the QSAR-context, it is necessary and sufficient that ρ verifies conditions (2), i.e., to be a dissimilarity index. In this sense MSD was used as a steric parameter (instead of the Taft E_S constant, for example) in Hansch type structure-biological activity correlations. 2) These results obtained for MSD are valid also for the Monte Carlo version (MCD) and for the optimized, topological version (MTD) of this method.[284]

8   CONCLUSIONS

The work described here is certainly tributary to some general difficulties of QSAR : if biological affinity depends on a large number of structural features, as it often does, one needs a corresponding number of predictor variables. If the number of experimental points is not several times larger, the reliability of the model (regressional equation) obtained is low, perhaps even if most of the challenged predictor variables are finally eliminated because of low significance. The use of test series, with correlation of experimental activities vs. calculated ones by means of the regressional equation determined for a different series of compounds, may be a good auxiliary in determining the real predictive value of the model.
Topological indices have been comparatively rarely used for QSAR, but the results obtained so far are promising. Further QSAR-studies with topological indices and especially development of new ones, characterizing quality and other steric features are badly needed.

The minimal steric difference concept allows a detailed description of steric structure - at about 1A precision, and is thus suited for description of steric fit. This precision may be nevertheless insufficient in certain instances, especially if imperfections of the superposition process are concerned. The Monte Carlo version (MCD) allows a practically unlimited precision in determining non-overlapping volumes. The part of the receptor which interacts within the studied series of molecules can be mapped by the MTD-procedure, but this requires a large number of predictor variables (the ε_j's).

Experimental vs. calculated activity correlations for test series (Chap.5) in MTD-studies, as well as simple A vs. MSD correlations with the "best" molecule of the series as standard, suggest that steric misfit generally accounts for about 50% of biological activity variance. One should not forget that some test results of § 5.3 and 5.4 suggest that additive effects of all types may account for only about 70% of this variance. The MTD-procedure accounts also for nonadditive steric fit effects, produced by the possibility of flexible molecules to adopt several low energy conformations, or to ·occupy different positions within the receptor. Although results with test series for the MTD-procedure demonstrate a statistical significance above 95% for the resulted receptor map as a whole, one must be careful about the individual ε_j-vertex assignments. One may thus expect the MTD procedure to be useful in predicting best molecular shapes for drug design.

Biological activity may often be determined by compartimentalization, passive transport phenomena and even receptors may have sometimes a low rigidity[55]. Recent results on conformational flexibility of proteins indicate a very wide range of flexibilities[285] but certainly steric fit is important for the majority of receptor-effector interactions. The value of the regressional coefficient (β) in MSD and MTD-studies ($\beta \sim 0.6$) indicates about 0.85 kcal/mol per one atom (nonhydrogen) steric misfit.

Like other QSAR procedures, those described here will have their range of applicability. Certainly, for series of molecules differing by rather small substituents in different positions of a molecular core, one should apply regressional equations with hydrophobicity and electronic parameters as described in the treatise of Purcell, Bass, and Clayton[3]. Steric effects for such series will be best described by the L, B_1-B_4 parameters of Verloop et al.[4]. Also, for series of molecules differing by a wide variety of large, flexible

and complex substituents, the use of indicator variables of the type
introduced by Hansch[85] is the best procedure. The proper domain of
MTD will be especially series of molecules with widely differing spa-
tial structure,e.g.stereoisomers, but covering not too different spa-
tial regions and without excessive flexibility. If the molecules of
the series present steric differences in several positions, substituted oligopeptides as example, the large number of predictor variables required will make the MTD-procedure quite unreliable. In such
cases, the simple MSD-procedure with the natural effector as standard
may give a first insight in the steric requirements of biological activity. Receptor characteristics will also determine the type of
approach to be used. Rigid receptors, where small steric features determine large affinity variations will require minimal steric difference techniques for QSAR, probably even those of the Monte Carlo
(MCD)-procedure.

One may finally observe that the description of steric structure via hypermolecule and x_{ij}-vectors may be used also in pattern recognition procedures.

9 APPENDIX

The appendix presents the two FORTRAN programs which implement the optimized MTD-procedure and the MCD-Monte Carlo procedure. The programs correspond to the methods described in the respective chapters(Chap. 5 and 6).

9.1. The MTD/1-program

The structure of this program is as follows

Subroutine	Function
MSDI	Calculates MTD values for the N molecules of the considered set
DYE	DETERMINES the hypermolecule vertex at which the attached ε-value is modified (i.e., $\varepsilon_{old} \longrightarrow \varepsilon_{new}$)
ORDON	Orders into an increasing series the ΔY values corresponding to $\varepsilon = +1$, 0 and -1.
RESOL	Calculates regressional coefficients by the least squares method.
SCR	Displays computational results
TITLU	Prints the title of the data set considered in computations.
CEND	Ensures reading of initial input data
CREST	Read entrase data Y_I, I=1,2,...N and the initial map(standard)[x)

The logic scheme of the MTD/1 program is given in Figure. 19

The input data for MTD/1 have the following structure:·

[x) Here, Y_I stands for A_I of Chap. 5, I for i

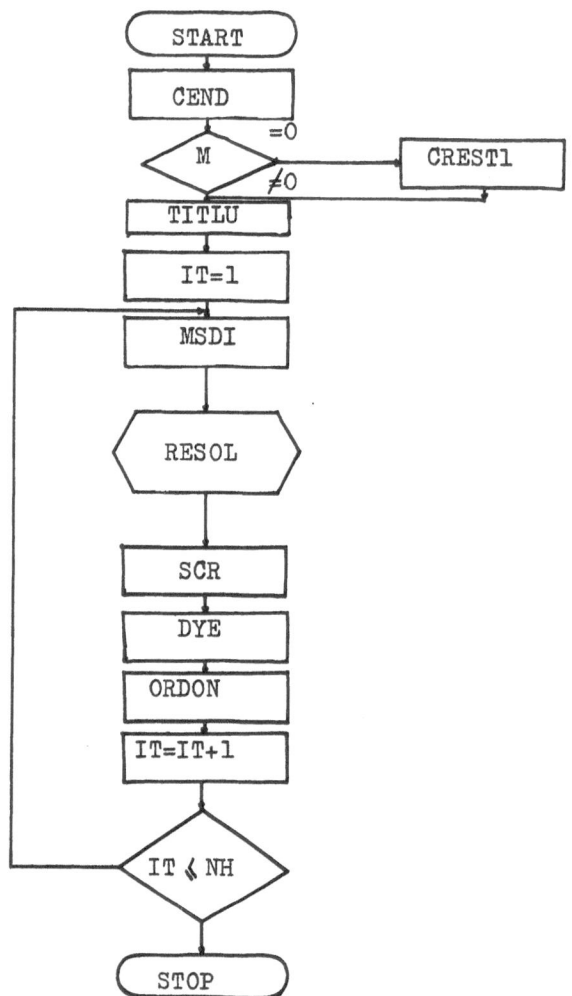

Figure 19   MTD/1 - flow chart

```
      READ(1o5,9oo)NAME,IEU

9oo FORMAT(20A1,19X,I1)

      READ(105, 905)N,M,NH

905 FORMAT(3I3)

      DO 2 I=1,N

  2 READ(105,910) (A(I), (SIG(J,I), J=1,M

910 FORMAT(8F10.4)

      DO 3 I=1,N

  3 READ(105,920) (X(I,J), J=1,NH)

920 FORMAT(40I1)

      READ(1o5,909)(E(J), J=1,NH)

909 FORMAT(40I2)
```

The identificators used hereby have the following significances :

Identificators	Significance
NAME	Informations concerning the data set
IEU	IEU=0, Y values are not logarithmized;
	IEU=1, Y values are logarithmized
N	Numbers of considered Y values
M	Number of predictor variables in the li-
	near model
NH[xx]	Number of vertexes of the hypermolecule
A	Values of the dependent variables Y_I
SIG	Values of the NxM predictor variables
	(M on one card)
X	$\mathcal{E}_i$ -values for a molecule(structure des-

[x*)]
   NH stands for M of Chap. 5

$$\text{criptor} \begin{bmatrix} \textbf{x}_{ij} \end{bmatrix} \text{)}$$

E   Start map (standard) with NH-vertices

<u>Observations</u>. Some defficiencies of the MTD method will be re-
moved in the MTD/2 version by reformulating the method within the
frame of graph theory (see refs. [286-288]).

The complete listing of the MTD/1 program is given below

```
C   PROGRAM  MTD/1
      INTEGER X(97,40),E(40)
      DIMENSION DY(40,3),MSD(97)
      DIMENSION SIG(3,100),A(100),NAME(20),AC(100)
      DIMENSION C(121),T(11),SIGB(10),DIFS(10)
    1 CALL CEND(NAME,IEU,&999,N,M,NH)
      IF(M.EQ.0)GO TO 120
      CALL CREST(A,SIG,M,N,X,NH,E,IEU)
  140 CONTINUE
      CALL TITLE(N,NH)
      DO 888 IT=1,NH
      IF(IT.GT.1)M=M-1
      CALL MSDI(X,E,SIG,MSD,N,M,NH)
      IM=M+1
      MI=IM**2
      DO 5 I=1,IM
    5 T(I)=0.
      DO 6 I=1,MI
```

```
6 C(I)=0.

  NG=M+1

  DO 7 J=1,M

  SIGB(J)=0.

  DIFS(J)=0.

7 CONTINUE

  CE=0.

  EPS=1.E-10

  C(1)=FLOAT(N)

  DO 10 J=1,M

  DO 10 I=1,N

10 C(J+1)=SIG(J,I)+C(J+1)

  KK=M+2

  C(KK)=C(2)

  DO 20 K=1,M

  DO 30 J=1,M

  DO 40 I=1,N

  CE=SIG(K,I)*SIG(J,I)+CE

40 CONTINUE

  KK=KK+1

  C(KK)=CE

  CE=0.

30 CONTINUE

  KK=KK+1

  C(KK)=C(K+2)

20 CONTINUE
```

```
      DO 50 I=1,N
50 T(1)=T(1)+A(I)
      DO 60 L=1,M
      DO 60 I=1,N
60 T(L+1)=T(L+1+SIG(L,I)*A(I)
      CALL RESOL(C,T,NG,KOD,EPS)
      IF(KOD.EQ.1)STOP
      BETA=-T(NG)
      AB=0.
      DO 70 I=1,N
      AC(I)=T(1)
      DO 80 J=1,M
      AC(I)=AC(I)+SIG(J,I)*T(J+1)
80 CONTINUE
      AB=AB+AC(I)
70 CONTINUE
      ABC=AB/N
      AB=0.
      DO 90 I=1,N
90 AB=AB+A(I)
      ABE=AB/N
      R1=0.
      R2=0.
      R3=0.
      DO 100 I=1,N
      R1=R1+(A(I)-ABE)*(AC(I)-ABC)
```

```
       R2=R2+(A(I)-ABE)**2

       R3=R3+(AC(I)-ABC)**2

100 CONTINUE

       R=R1/SQRT(R2*R3)

       CALL SCR(MSD,N,E,NH,R,IT)

       CALL DYE(X,E,DY,N,NH,A,AC,BETA)

       CALL ORDON(DY,NH,I,K,&111)

       E(I)=K

888 CONTINUE

111 CALL FIN(T,NG,DY,NH)

       GO TO 1

999 STOP

120 CONTINUE

       CALL CRESTI(A,SIG,M,N,X,NH,E,IEU)

       GO TO 140

       END

       SOUBROUTINE MSDI(X,E,SIG,MSD,N,M,NH)

       INTEGER X(97,40),E(40)

       INTEGER S

       DIMENSION SIG(3,100),MSD(97)

       S=0

       DO 1 J=1,NH

       IF(E(J).EQ.-1)S=S+1

  1 CONTINUE

       DO 3 I=1,N

       MSD(I)=S
```

```
      DO 2 J=1,NH
  2 MSD(I)=MSD(I)+E(J)*X(I,J)
      SIG(M+1,I)=FLOAT(MSD(I))
  3 CONTINUE
      M=M+1
      RETURN
      END
      SUBROUTINE DYE(X,E,DY,N,NH,A,AC,BETA)
      INTEGER X(97,40),E(40)
      DIMENSION DY(40,3),A(100),AC(100)
      DO 3 J=1,NH
      NO=0
      N1=0
      DO 1 I=1,N
      IF(X(I,J).EQ.1)N1=N1+1
      IF(X(I,J).EQ.O)NO=NO+1
  1 CONTINUE
      S=0
      DO 2 I=1,N
      IF(X(I,J).EQ.1)S=S+A(I)-AC(I)
  2 CONTINUE
      DY(J,1)=2*BETA*(-1-E(J)*S+BETA**2*NO*N1*(-1-E(J))**2/N
      DY(J,2)=2*BETA*( O-E(J)*S+BETA**2*NO*N1*( O-E(J))**2/N
      DY(J,3)=2*BETA*( 1-E(J)*S+BETA**2*NO*N1*( 1-E(J))**2/N
  3 CONTINUE
      RETURN
```

```
      END

      SUBROUTINE ORDON(DY,NH,I,K,*)

      DIMENSION DY(40,3)

      I1=1

      I2=1

      I3=1

      DO 1 I=1,NH

      IF(DY(I,1).LT.DY(I1,1))I1=I

      IF(DY(I,2).LT.DY(I2,2))I2=I

      IF(DY(I,3).LT.DY(I3,3))I3=I

    1 CONTINUE

      IF(DY(I1,1).GE.O..AND.

     1    DY(I2,2).GE.O..AND.

     2    DY(I3,3).GE.O.)RETURN 1

      IF(DY(I1,1).LE.DY(I2,2).AND.DY(I1,1).LE.DY(I3,3))GO TO 2

      IF(DY(I2,2).LE.DY(I1,1).AND.DY(I2,2).LE.DY(I3,3))GO TO 3

      IF(DY(I3,3).LE.DY(I1,1).AND.DY(I3,3).LE.DY(I2,2))GO TO 4

      RETURN 1

    2 I=I1

      K=-1

      RETURN

    3 I=I2

      K=0

      RETURN

    4 I=I3

      K=1
```

```
      RETURN

      END

      SUBROUTINE SCR(MSD,N,E,NH,R,IT)

      INTEGER E(40)

      DIMENSION MSD(97)

      WRITE(108,911)

      WRITE(108,902)IT,(MSD(I),I=1,N)

      WRITE(108,903)(E(J),J=1,NH)

      WRITE(108,904)R

      RETURN
  911 FORMAT(1X,'I     I ',61X,',50X,' I',' ',7X,' I')
  902 FORMAT(1X,'I    ',I2,'I ',20I3/4(1X,'I ',2X,'I ',20I3,'I',150,'
     I'/))
  903 FORMAT('+',67X,'I ',50I1,'I')
  904 FORMAT('+',121X,'I',F7.5,'I')
      END

      SUBROUTINE FIN(T,NG,DY,NH)

      DIMENSION T(11),DY(40,3)

      WRITE(108,980)T(1)?(I,T(I),I=2,NG)

      WRITE(1o8,906)

      DO 110 J=1,3

      K=-2+J

      WRITE(108,907)K,(DY(I,J),I=1,NH)

      WRITE(108,906)
  110 CONTINUE

      RETURN
```

```
980 FORMAT(5X,'ROOTS OF EQUATION='/

  C       5X,'--------------------'/

  C       5X,'ALPHA    =',F15.10/

  C   100(5X,'BETA(', I2,')=',F15.10/))

906 FORMAT(1X,131('-'))

907 FORMAT(1X,'I DY(',I2,') ',25F5.1/1X,'I            ',25F5.1

    END

    SUBROUTINE TITLE(N,NH)

    WRITE(108,901)N,NH

    RETURN

901 FORMAT(1X,131('-')/

  1 ' ','* N *    ',20X,'MSD(I),I=1,2,3,.....',I2,20X,

  2'*',17X,'E(I),I=1,2,3,...',I2,17X,'*',3X,'C.C.',2X,'*'/

  31X,131('-'))

    END

    SUBROUTINE NU(NAME)

    DIMENSION NAME(20)

    WRITE(108,930)(NAME(IJ),IJ=1,20)

    RETURN

930 FORMAT(20A1)

    END

    SUBROUTINE CEND(NAME,IEU,*,N,M,NH)

    DIMENSION NAME(20)

    READ(105,900,END=999)(NAME(IJ),IJ=1,20),IEU

    READ(105,905)N,M,NH

    RETURN
```

```
999 RETURN 1

900 FORMAT(20A1,19X,I1)

905 FORMAT(3I3)

    END

    SUBROUTINE CREST(A,SIG,M,N,X,NH,E,IEU)

    DIMENSION SIG(3,100),A(100)

    INTEGER X(97,40),E(40)D

    DO 2 I=1,N

    READ(1o5,910)(A(I),(SIG(J,I),J=1,M))

    IF(IEU.EQ.1)A(I)=ALOG10(A(I))

  2 CONTINUE

140 CONTINUE

    DO 3 I=1,N

    READ(105,920)(X(I,J),J=1,NH)

  3 CONTINUE

    READ(105,909)(E(J),J=1,NH)

    RETURN

    ENTRY CREST1(A,SIG,M,N,X,NH,E,IEU)

    DO 130 I=1,N

130 READ(105,910)A(I)

    GO TO 140

909 FORMAT(40I2)

910 FORMAT(8F10.4)

920 FORMAT(40I1)

    END
```

9.2   The MCD-Program

The MCD-Program corresponds to the method described in Chap.6

The random number generator used here is ALEAT, of the mathematical library of the IRIS-50 computer. The MCD-program is described below and the complete listing is reproduced. The subroutine for computation of cartesian coordinates was realized by adaptation of the program written by Ray.[286]

The interaction of the subroutines is illustrated in Figure 20.

9.2.1   Succession of Computation Steps in the Program

The steps of the computing procedure which allow to obtain the MCD values are:

1. Computation of the atomic cartesian coordinates of the considered molecules.

2. Establishing of the standard molecule,S. The most active molecule of the considered series is considered as standard.

3. The atomic coordinates of each molecule are memorized on the magnetinc band.

4. The minimal parallelipipedum is determined which circumscribes the van der Waals envelopes of the (superposed) pairs standard/each molecule of the series.

5. The MCD values attached to each pair mentioned at point 4 are calculated according to the procedure described at Chap. 6.

6. The regressional computations are performed in two steps:

a) A_i vs. $1, \ldots, k$

b) A_i vs. $1, \ldots, k$, MCD

$_I$, I=1,...,k represent predictor variables expressing nonsteric effects.

9.2.2 Description of the Subroutines Used

1.	VAN WAALS	Determines van der Waals radii
2.	CORELIN	Subprogram of linear correlation
3.	TABEL	Initialisation of values for the Fischer-test
4.	INPUT1	Executes reading of the input data for CORELIN
5.	STD	Establishes the standard molecule
6.	RESOL	Subprograme for solving a system of linear equations
7.	R2	Assigns working areals for CORELIN
8.	CORD	Calculates coordinates of molecules which enter in the calculation of MCD
9.	INPUT2	Executes reading of input data for CORD
10.	OUT	Executes printing of input and output data for CORD
11.	BUFOUTMT	Memorates on magnetic band intermediate data resulted out of the computations executed in CORD
12.	MATP	Calculates interatomic distances for CORD
13.	BUFINPMT	Reads of the band data memorized by BUFOUTMT
14.	MAX	Calculates dimensions and coordinates of the parallelipipedum which circumscribes a pair of molecules
15.	MSD	Calculates MCD values

16. UNIREL   Generator of numbers uniformly distributed on
                     the(a,b) interval

17. ALEAT   Generator of numbers uniformly distributed
                     on the(0,1) interval

18. OUTMSD   Prints MCD values

19. MODIF   Reinitialises data for CORELIN

20. R1   Assigns working areas for CORD

9.2.3 Standard Subroutines Used

1. SQRT   Calculates square root

2. FLOAT   Integer-real conversion

3. ARS   Calculates absolutes values

4. COS,SIN   Executes trigonometric sin and cos functions

5. BUFFER OUT Transfer a memory zone on magnetic band

6. BUFFER IN Transfer a magnetic band block on a memory
                zone

7. AMAX1   Determines the maximum value among two or more
               real values

8. AMIN1   Determines the minimal value among two or more
               real values

9. ALOGIO   Executes the logaritmic function in basis 10 of
               a number.

9.2.4 Card Indexes

NAME	INDEX	SUPORT	ARTICLE FORMAT
1. SALV	A	magnetic band	blocked
2. REST	B	magnetic band	blocked

Observation. The two card indexes are used by the standard subprograms BUFFER IN and BUFFER OUT.

9.2.5 Input Data

Card 1

```
        READ(105,900)N,M

900     FORMAT(2I3)
```

                N - number of molecules in the set

                M - number of SIGMA parameters

Card 2↓N

```
        DO  10 I=1,X1

10    READ(105,905) (A(I),(SIG(J,I),J=1,M)

905   FORMAT (8F10.4)
```

              A- experimental Y_I values

              SIG- SIGMA parameters

Card N+1 ....

```
        READ(105,900,END=999) (NAME(I),I=1,20)

        READ(105,910)NOAT,(IZAT(I),I=1,3)KWIK,R13,R23,THETA

        READ(105,930)NA,NB,NC,ND,IZAT(ND),ILAZY RCD,THBCD,PABCD

900   FORMAT (20A4)

910   FORMAT(5I3,2F7.4,F14.7)

930   FORMAT(6I3,F7.4,2F14.7)
```

Data which characterize a molecule. Significance of each variable corresponds to the CORD program registred at QCPE of Indiana
University under the number 226.

    There exists N sets for each molecule per se.

    After the last set one puts an EOF card.

9.2.6 Structure of Output Data

Output data are divided into three classes which appear suc-cesively, as follows:

- coordinates

- MCD values

- multilinear regression resupts

Coordinates

Contains the input data required to obtain the coordinates,the catesian coordinates and interatomic distances for each molecule of the set.

MCD values

Contains the number of molecules in the set and the MCD values calculated with respect to the standard. The standard molecule is the one for which the calculated MCD=0.

Multilinear regression resuls

Prints partial correlation coefficients, the correlation coefi-cient, F-test value in two variants, with and without MCD.

9.2.7 Error Messages

Error type	Error description
1. **ERROR**	incident produced in BUFFER IN or in BUFFER OUT
2. **ERROR** NUMBER OF MOLECULES HIGHER THAN THE VALUE CONSIDERED AT THE BEGINNING	The value of the N-parameter is lower or higher than the number of molecu-les in the data set.

Restrictions

1. The number of molecules in the set and of the SIGMA parameters

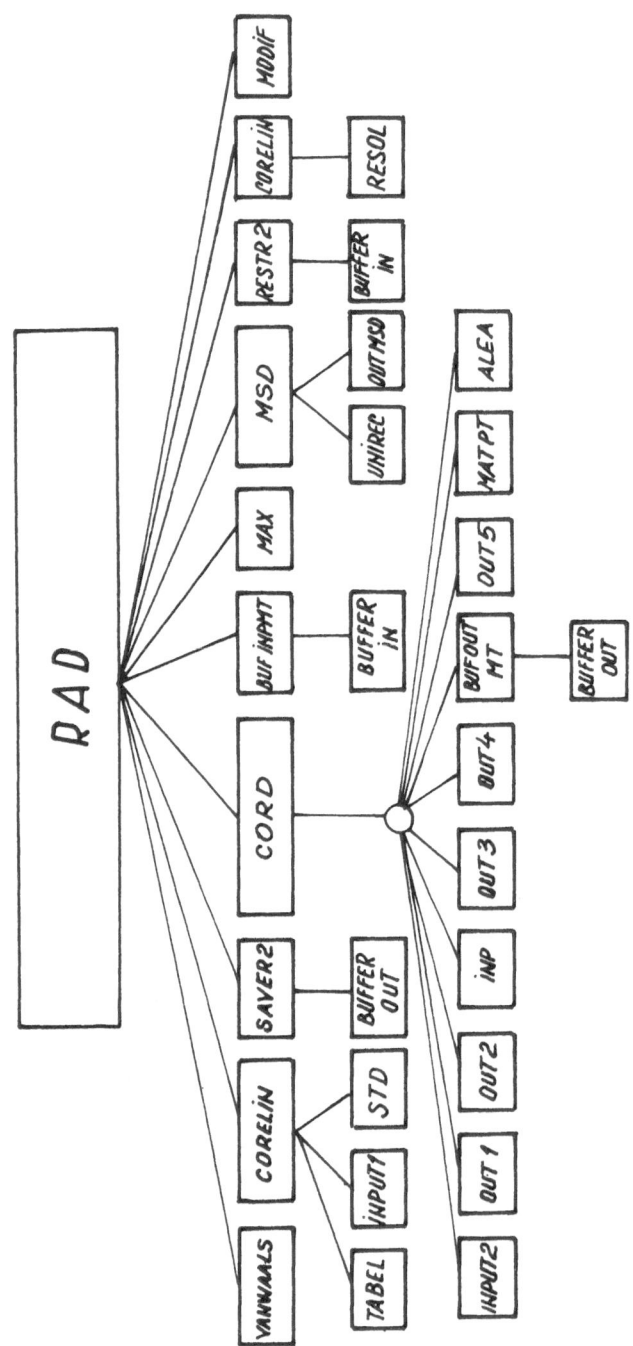

Figure 20: Interaction of subroutines in the MCD-program.

depends on the size of tabels. In this variant the maximal values

are, respectively, 50 molecules of at most 20 atoms each and 20

SIGMA parameters.

2. For computers differing from IRIS 50 the MASKEP subprogram is

not obligative.

3. The minimal memory required by this program is 90 KO, for the

situation in which the segmented program version is maintained.

The listing of this program is:

```
C    PROGRAM MCD
* DEFINE FILE AO(DVT:MT,RCF,B)=1
* DEFINE FILE BO(DVT:MT,RCF,B)=2
      COMMON VOL,NI,NT
      COMMON/TMSD/TMSD(100)
      COMMON/GRAD/M,N
      COMMON/STANDARD/MS,NATS,IZAS(30),XS(30),YS(30),ZS(30)
      COMMON/VANDER/WAALS(100),NRW
      CALL VANWAALS
      CALL CORELIN
      CALL SAVER2
      CALL CORD
    1 CALL BUFINPMT(&2)
      CALL MAX
      CALL MSD
      GO TO 1
```

```fortran
2 CALL RESTR2

  CALL CORELIN1

  CALL MODIF

  CALL CORELIN1

  STOP

  END

  SUBROUTINE UNIREL(A,B,Z,IU,X)

  CALL MAS

  CALL ALEAT(IU,IB,X)

  CALL NMAS

  IU=IB

  ENTERY UNIRELI(A,B,Z,X)

  Z=A+(B-A)*X

  RETURN

  END

  SUBROUTINE ALEAT(IA,IB,Z)

  IB=IA*65539

  IF(IB.GE.O)GO TO 1

  IB=IB+2147483647+1

1 Z=IB

  Z=Z*0.4656613E-9

  RETURN

  END

  SUBROUTINE R1

  COMMON/LIMIT/XMAXP,YMAXP,ZMAXP,XMAXN,YMAXN,ZMAXN

  COMMON/CURENT/SW,NATC,IZAT(50),X(50),Y(50),Z(50)
```

```
       COMMON /NAME/NAME(20)

       COMMON/COORD/NA,NB,NC,ND,ILAZY,RCD,THBCD,PABCD

       COMMON/CORDU/T(90,90),NOAT,KWIK,R12,R23,THETA

       COMMON/RANDOM/I1,I2,S,J1,J2,P,K1,K2,Q

       RETURN

       SUBROUTINE R2

       COMMON/GAP/SIG(20,&0),SIGB(20),SIGR(20,50)

       COMMON/EXP/A(100),C(441),T(21),AC(100)

       COMMON/GRLIB/TAB(2,27),ITAB(20),DIFS(20),RRR(20,20),RR(20)

       ENTERY SAVER2

       CALL BUFFER OUT(2,1,SIG,3196,II,III)

       IF(II.EQ.4)WRITE(108,900)

       REWIND 2

       RETURN

       ENTRY RESTR2

       CALL BUFFER IN(2,1,SIG,3196,II,III)

       IF(II.EQ.4)WRITE(108,900)

   900 FORMAT(///10X, **ERROR** '')

       RETURN

       END

       SUBROUTINE CORELIN

       INTEGER SW

       COMMON/GAP/SIG(20,50),SIGB(20),SIGR(20,50)

       COMMON/EXP/A(100),C(441),T(21),AC(100)

       COMMON/GRLIB/TAB(2,27),ITAB(20),DIFS(20),RRR(20,20),RR(20)

       COMMON/GRAD/M,N
```

```
      CALL TABEL

      CALL INPUT1

      CALL STD

      RETURN

      ENTRY CORELIN1

      INTEGER SW

      SW=1

4 CONTINUE

      WRITE(108,935)

      DO 3 I=1,N

3 WRITE(108,940)(I,A(I),(SIG(J,I),J=1,M))

      IM=M+1

      MI=IM**2

      DO 5 I=1,IM

5 T(I)=0.

      DO 6 I=1,MI

6 C(I)=0.

      NG=M+1

      DO 7 J=1,M

      SIGB(J)=0.

      DIFS(J)=0.

7 CONTINUE

      DO 8 K=1,M

      DO 8 J=1,M

      RRR(J,K)=0.

8 CONTINUE
```

```
      CE=0.

      EPS=1.E-10

      C(1)=FLOAT(N)

      DO 10 J=1,M

      DO 10 I=1,N

   10 C(J+1)=SIG(J,I)+C(J+1)

      KK=M+2

      C(KK)=C(2)

      DO 20 K=1,M

      DO 30 J=1,M

      DO 40 I=1,N

      CE=SIG(K,I)*SIG(J,I)+CE

   40 CONTINUE

      KK=KK+1

      C(KK)=CE

      CE=0.

   30 CONTINUE

      KK=KK+1

      C(KK)=C(K+2)

   20 CONTINUE

      DO 50 I=1,N

   50 T(1)=T(1)+A(I)

      DO 60 L=1,M

      DO 60 I=1,N

   60 T(L+1)=T(L+1)+SIG(L,I)*A(I)

      CALL RESOL(C,T,NG,KOD,EPS)
```

```
      WRITE(108,945)KOD
      WRITE(108,980)T(1),(I,T(I),I=2,NG)
      AB=0.
      DO 70 I=1,N
      AC(I)=T(1)
      DO 80 J=1,M
      AC(I)=AC(I)+SIG(J,I)*T(J+1)
   80 CONTINUE
      AB=AB+AC(I)
   70 CONTINUE
      ABC=AB/N
      AB=0.
      DO 90 I=1,N
   90 AB=AB+A(I)
      ABE=AB/N
      R1=0.
      R2=0.
      R3=0.
      WRITE(108,915)
      WRITE(108,955)(I,AC(I),I=1,N)
      WRITE(108,925)ABC,ABE
      DO 100 I=1,N
      R1=R1+(A(I)-ABE)*(AC(I)-ABC)
      R2=R2+(A(I)-ABE)**2
      R3=R3+(AC(I)-ABC)**2
  100 CONTINUE
```

```
      R=R1/SQRT(R2*R3)
      WRITE(108,950)R
      ADIF=0.
      DO 110 I=1,N
      ADIF=ADIF+(AC(I)-A(I))**2
110   CONTINUE
      S=SQRT(ADIF)
      S=S/SQRT(FLOAT(N-1))
      WRITE(108,960)S
      DO 130 J=1,M
      DO 120 I=1,N
      SIGB(J)=SIGB(J)+SIG(J,I)
130   CONTINUE
      SIGB(J)=SIGB(J)/N
130   CONTINUE
      DO 150 J=1,M
      DO 140 I=1,N
      DIFS(J)=DIFS(J)+(SIG(J,I)-SIGB(J))**2
140   CONTINUE
150   CONTINUE
      DO 170 J=1,M
      DIF1=0.
      DIF2=0.
      DO 160 I=1,N
      DIF1=DIF1+(A(I)-ABE)*(SIG(J,I)-SIGB(J))
      DIF2=DIF2+(A(I)-ABE)**2
```

```
160 CONTINUE

    DIF2=DIF2*DIFS(J)

    RR(J)=DIF1/SQRT(DIF2)

170 CONTINUE

    WRITE(108,965)(RR(J),J=1,M)

    DO 190 K=1,M

    DO 190 J=1,M

    DDIF=0.

    DO 180 I=1,N

    DDIF=DDIF+(SIG(J,I)-SIGB(J))*(SIG(K,I)-SIGB(K))

180 CONTINUE

    RRR(J,K)=DDIF/SQRT(DIFS(J)*DIFS(K))

190 CONTINUE

    WRITE(108,970)((RRR(J,K),K=1,M,J=1,M)

    WRITE(108,975)

    IF(R.EQ.1.) GO TO 1100

    F=(FLOAT(N-M+1)-1)*R*R)/(FLOAT(M+1)*(1-R*R))

    WRITE(108,995)F

1100 A1=0.

    A2=0.

    DO 195 I=1,N

    A1=A1+(AC(I)-A(I))**2

    A2=A2+(A(I)-ABE)**2

195 CONTINUE

    IF(N.EQ.(M+1)) GO TO 1200

    EV=1.-((A1/FLOAT(N-M+1))))/(A2/FLOAT(N-1))
```

```
      WRITE(108,996)EV
1200 CONTINUE
      GO TO (200,1),SW
 200 GRLIB=FLOAT(N-2))
      DO 210 L=1,27
      IF(GRLIB.LE.TAB(1,L))GO TO 220
 210 CONTINUE
 220 L=L+1
      RMIN=TAB(2,L-1)
      IL=0.
      DO 250 J=1,M
      IF(ABS(RR(J)).LT.RMIN)GO TO 240
      IL=IL+1
      ITAB(IL)=J
      DO 230 I=1,N
 230 SIGR(IL,I)=SIG(J,I)
 240 CONTINUE
 250 CONTINUE
      IF(IL.GT.0)GO TO 260
      WRITE(108,920)
      WRITE(108,985)GRLIB,RMIN,(RR(J),J=1,M)
      GO TO 1
 260 CONTINUE
      M=IL
      DO 270 J=1,M
      DO 270 I=1,N
```

```
270 SIG(J,I)=SIGR(J,I)

    WRITE(108,990)(J,ITAB(J),J=1,M)

    SW=2

    GO TO 4

  1 RERURN

915 FORMAT(//10X,CALCULATED A VALUES ARE://10X,6('MOLECULE A CALC.

    C ')//)

920 FORMAT(////'*TOTI R(J) RMIN*'//)

925 FORMAT(//10X,'A BARED CALC.:',F14.7,5X,'A BARED EXP.:',F14.7/)

935 FORMAT(//3X,'MOLECULE',2X,'A  EXP.',3X,'SIGMA 1',4X,'SIGMA 2',

    C4X,'SIGMA 3',4X,'SIGMA 4',4X,'SIGMA 5',4X,'SIGMA 6',4X,

    C'SIGMA 7',4X,'SIGMA 8',4X,'SIGMA 9',4X,'SIGMA 10'///)

940 FORMAT(6X,10(I3,3X,F9.3))

945 FORMAT(//10X,'KOD=',I3///)

950 FORMAT(//2X,130('*')///10X,'REZULTS:',5X,

    C'CORRELATION COEFFICIENT  ≐  ',F8.5///)

955 FORMAT(/10X,I6,F14.6,I6,F14.6,I6,F14.6,I6,F14.6,I6,F14.6,

    1            I6,F14.6)

960 FORMAT(//10X,'COEFFICIENT S = ',F11.6)

965 FORMAT(//10X,'PARTIAL CORRELATION COEFFICIENTS E(J)=',

    1    ///(5X,12F10.6))

970 FORMAT(//10X,'CORRELATION COEFFICIENS R(J,K)=',

    1    ///(10X,7F13.6))

975 FORMAT(//2X,'OBS.: R(1,1)...R(1,M),R(2,1),...,R(2,M),...R(M,M

    1)')

980 FORMAT(5X,'ROOTS OF EQUATION = '/
```

```
C       5X,'------------------ '/
C       5X,'ALPHA =',F15.10/
C  100(5X,'BETA(',I2,')=',F15.10/))
985 FORMAT(//5X,'GR. LIB.=',F5.0,5X,'R MIN.=',F6.3/24X,13('x')5X//,
C           'R(J)(J=1,M)=',5X,18F6.3)
990 FORMAT(///5X,'PARTIAL CORRELATION'/5X,17('x')/
C        10(5X,2('SIGMA(',I2,')=SIGMA(',I2,')')/))
995 FORMAT(10X,'xxTEST Fxx'/10X,10('-')////10X,'F=',F13.6////)
996 FORMAT(///10X,'EV=',F13.6/////)
    END
    SUBROUTINE RESOL(A,B,KOD,EPS)
    DIMENSION A(441),B(21)
    KOD=0.
    JJ=-N
    DO 8 J=1,N
    JY=J+1
    JJ=JJ+N+1
    EMAX=0.
    IT=JJ-J
    DO 2 I=J,N
    IJ=IT+I
    IF(ABS(EMAX)-ABS(A(IJ)))1,2,2
  1 EMAX=A(IJ)
    IND=I
  2 CONTINUE
    IF(ABS(EMAX)-EPS) 3,3,4
```

```
3 KOD=1

  RETURN

4 II=J+N*(J-2)

  IT=IND-J

  DO 5 K=J,N

  II=II+N

  IJ=II+IT

  R=A(II)

  A(II)=A(IJ)

  A(IJ)=R

  A(II)=A(II)/EMAX

5 CONTINUE

  R=B(IND)

  B(IND)=B(J)

  B(J)=R/EMAX

  IF(J-N)6,9,6

6 IO=N*(J-1)

  DO 8 IK=JY,N

  IKK=IO+IK

  IT=J-IK

  DO 7 JK=JY,N

  IKJ=N*(JK-1)+IK

  JKK=IKJ+IT

7 A(IKJ)=A(IKJ)-A(IKK)*A(JKK)

8 B(IK)=B(IK)-B(J)*A(IKK)

9 N1=N-1
```

```
      IT=N*N

      DO 10 J=1,N1

      IA=IT-J

      IB=N-J

      IC=N

      DO 10 K=1,J

      B(IB)=B(IB)-A(IA)*B(IC)

      IA=IA-N

      IC=IC-1

10    CONTINUE

      RETURN

      END

      SUBROUTINE TABEL

      COMMON/GRLIB/TAB(2,27),ITAB(20),DIFS(20),RRR(20,20),RR(20)

      DO 10 I=1,20

10    TAB(1,I)=FLOAT(I)

      TAB(1,27)=100.

      TAB(1,26)=80.

      TAB(1,25)=60.

      TAB(1,24)=50.

      TAB(1,23)=40.

      TAB(1,22)=30.

      TAB(1,21)=25.

      TAB(2,1)=0.997

      TAB(2,2)=0.950

      TAB(2,3)=0.878
```

```
TAB(2,4)=0.811

TAB(2,5)=0.754

TAB(2,6)=0.707

TAB(2,7)=0.666

TAB(2,8)=0.632

TAB(2,9)=0.602

TAB(2,10)=0.576

TAB(2,11)=0.553

TAB(2,12)=0.532

TAB(2,13)=0.514

TAB(2,14)=0.497

TAB(2,15)=0.482

TAB(2,16)=0.468

TAB(2,17)=0.456

TAB(2,18)=0.444

TAB(2,19)=0.433

TAB(2,20)=0.423

TAB(2,21)=0.381

TAB(2,22)=0.349

TAB(2,23)=0.307

TAB(2,24)=0.273

TAB(2,25)=0.250

TAB(2,26)=0.217

TAB(2,27)=0.195

RETURN

END
```

```
      SUBROUTINE STD

      COMMON/EXP/A(100),C(441),T(21),AC(100)

      COMMON/STANDARD/MS,NATS,IZAS(30),XS(30),YS(30),ZS(30)

      COMMON/GRAD/M,N

      MS=1

      DO 10 I=2,N

      IF(A(I).GT.A(MS))MS=I

   10 CONTINUE

      RETURN

      END

      SUBROUTINE INPUT1

      COMMON/GRAD/M,N

      COMMON/GAP/SIG(20,50),SIGB(20),SIGR(20,50)

      COMMON/EXP/A(100),C(441),T(21),AC(100)

      READ(105,900)N,M

      PRINT 900,N,M

      DO 10 I=1,N

      READ(105,905)(A(I),(SIG(J,I),J=1,M))

      A(I)=ALOG10(A(I))

  900 FORMAT(2I3)

  905 FORMAT(8F10.4)

   10 CONTINUE

      RETURN

      END

      SUBROUTINE CORD

      INTEGER SW
```

```
      COMMON VOL,NI,NT
      COMMON/CURENT/SW,NATC,IZAT(50),X(50),Y(50),Z(50)
      COMMON/NAME/NAME(20)
      COMMON/COORD/NA,NB,NC,ND,ILAZY,RCD,THBCD,PABCD
      COMMON/CORDU/T(90,90),NOAT,KWIK,R12,R23,THETA
      DIMENSION R(90,90)
      SW=0.
   99 CALL INPUT2(&999)
      CALL OUT1
      IF(KWIK-1)1,2,3
    1 CCOS=-1./3.
      SSIN=(2./3.)*SQRT(2.)
      GO TO 4
    2 CCOS=-0.5
      SSIN=0.5*SQRT(3.)
      GO TO 4
    3 THETA=THETA*3.1415926536/180.
      CCOS=COS(THETA)
      SSIN=SIN(THETA)
    4 DO 51 I=1,3
      X(I)=0.0
      Y(I)=0.0
   51 Z(I)=0.0
      X(2)=R12
      X(3)=R12-R23*CCOS
      Y(3)=R23*SSIN
```

```
CALL OUT2

DO 52 I=4,NOAT

CALL INP
CALL OUT3

IF(ILAZY-8)79,78,79

78 RBC=SQRT(X(NC)-X(NB))**2+(Y(NC)-Y(NB))**2+(Z(NC)-Z(NB))**2)

   X(ND)=X(NC)+(X(NC)-X(NB))*RCD/RBC

   Y(ND)=Y(NC)+(Y(NC)-Y(NB))*RCD/RBC

   Z(ND)=Z(NC)+(Z(NC)-Z(NB))*RCD/RBC

   GO TO 52

79 XA=X(NA)-X(NC)

   YA=Y(NA)-Y(NC)

   ZA=Z(NA)-Z(NC)

   XB=X(NB)-X(NC)

   YB=Y(NB)-Y(NC)

   ZB=Z(NB)-Z(NC)

   XYB=SQRT(XB**2+YB**2)

   K=1

   IF(XYB-0.1)9,10,10

 9 K=0

   XPA=ZA

   ZPA=-XA

   XA=XPA

   ZA=ZPA

   XPB=ZB

   ZPB=-XB
```

```
      XB=XPB

      ZB=ZPB

      XYB=SQRT(XB**2+YB**2)

   10 CONTINUE

      COSTH=XB/XYB

      SINTH=YB/XYB

      XPA=XA*COSTH+YA*SINTH

      YPA=YA*COSTH-XA*SINTH

      RBC=SQRT(XB**2+YB**2+ZB**2)

      SINPH=ZB/RBC

  105 COSPH=SQRT(1.-SINPH**2)

      XQA=XPA*COSPH+ZA*SINPH

      ZQA=ZA*COSPH-XPA*SINPH

      YZA=SQRT(YPA**2+ZQA**2)

      COSKH=YPA/YZA

      SINKH=ZQA/YZA

      IF(ILAZY-1)13,14,15

   13 COSD=1.0

      SIND=0.

      GO TO 21

   14 COSD=0.5

      SIND=0.5*SQRT(3.)

      GO TO 21

   15 IF(ILAZY-3)16,17,18

   16 COSD=-0.5

      SIND=0.5*SQRT(3.)
```

```
      GO TO 21

15 IF(ILAZY-3)16,17,18

16 COSD=-0.5

      SIND=0.5*SQRT(3.)

      GO TO 21

17 COSD=-1.0

      SIND=0.

      GO TO 21

18 IF(ILAZY-5)19,20,22

19 COSD=0.5

      SIND=-0.5*SQRT(3.)

      GO TO 21

20 COSD=0.5

      SIND=-0.5*SQRT(3.)

21 COSA=-1.0/3.0

      SINA=(2./3.)*SQRT(2.)

      GO TO 29

22 IF(ILAZY-7)23,24,26

23 COSD=1.0

      SIND=0.

      GO TO 25

24 COSD=-1.0

      SIND=0.

25 COSA=-0.5

      SINA=0.5*SQRT(3.)

      GO TO 29
```

```
26 IF(ILAZY-9)27,28,28

27 CONTINUE

   GO TO 29

28 THBCD=THBCD*3.1415926536/180.

   PABCD=PABCD*3.1415926536/180.

   SINA=SIN(THBCD)

   COSA=COS(THBCD)

   SIND=SIN(PABCD)

   COSD=COS(PABCD)

29 CONTINUE

   XD=RCD*COSA

   YD=RCD*SINA*COSD

   ZD=RCD*SINA*SIND

   YPD=YD*COSKH-ZD*SINKH

   ZPD=ZD*COSKH+YD*SINKH

   XPD=XD*COSPH-ZPD*SINPH

   ZQD=ZPD*COSPH+XD*SINPH

   XQD=XPD*COSTH-YPD*SINTH

   YQD=YPD*COSTH+XPD*SINTH

   IF(K-1)31,32,31

31 XRD=-ZQD

   ZRD=XQD

   XQD=XRD

   ZQD=ZRD

32 X(ND)=XQD+X(NC)

   Y(ND)=YQD+Y(NC)
```

```
       Z(ND)=ZQD+Z(NC)

52 CONTINUE

       CALL OUT4

       NATC=NOAT

       CALL BUFOUTMT

       DO 88 I=1,NOAT

       DO 88 J=1,NOAT

88 R(I,J)=SQRT(X(I)-X(J))**2+(Y(I)-Y(J))**2+(Z(I)-Z(J))**2)

       CALL OUT5

       CALL MATPT(NOAT,R)

       GO TO 99

999 REWIND 1

       RETURN

       END

       SUBROUTINE MATPT(N,A)

       COMMON VOL,NI,NT

       DIMENSION A(90,90)

       M=N

       MA=30

       NC=10

       II=0

       KK=0

       NCM1=NC-1

       J=0

       L=1

       IF(II-1)13,13,14
```

```
14 L=II-10

   II=0

   KK=5

13 DO 5 IZ=L?M,NC

   NIF=IZ+NCM1

   IF(NIF.GT.M)NIF=M

   J=J+N-IIx(IZ-1)

   IF(J-52)6,7,7

 7 I=0

   J=0

   GO TO 8

 6 I=1

 8 CONTINUE

   IF(I+KK-1)2,3,18

 2 WRITE(108,1)

 1 FORMAT(1H1)

 3 WRITE(108,4)(K,K=IZ,NIF)

 4 FORMAT(1H1,/10I10)

18 IJ=2x(NIF-IZ+1)+1

   IF(II)9,9,10

10 DO 11 IR=IZ,N

   JJ=IR

   IF(JJ.GT.NIF)JJ=NIF

11 WRITE(108,100)IR,(A(IR,IC),IC=IZ,JJ)

   GO TO 5

 9 DO 12 IR=1,N
```

```
   12 WRITE(108,100)IR,(A(IR,IC),IC=IZ,NIF)
  100 FORMAT(1H,I2,2X,10F10.4)
    5 CONTINUE
      RETURN
      END
      SUBROUTINE INPUT2(x)
      INTEGER SW
      COMMON/GRAD/M,N
      COMMON/CURENT/SW,NATC,IZAT(50),X(50),Y(50),Z(50)
      COMMON/NAME/NAME(20)
      COMMON/COORD/NA,NB,NC,ND,ILAZY,RCD,THBCD,PABCD
      COMMON/CORDU/T(90,90),NOAT KWIK,R12,R23,THETA
      READ(105,910)NOAT,(IZAT(I),I=1,3),KWIK,R12,R23,THETA
      SW=SW+1
      IF(3W.GT.N)WRITE(108,920)
      RETURN
      ENTRY INP
      READ(105,930)NA,NB,NC,ND,IZAT(ND),ILAZY,RCD,THBCD,PABCD
      RETURN
  999 RETURN 1
  900 FORMAT(20A4)
  910 FORMAT(5I3,2F7.4F14.7)
  920 FORMAT(///10X,`xxERRORxx NR. MOL. DEP. VAL. INIT. ')
  930 FORMAT(6I3,F7.4,2F14.7)
      END
      SUBROUTINE OUT
```

```
      COMMON/CURENT/SW,NATC,IZAT(50),X(50),Y(50),Z(50)

      COMMON/NAME/NAME(20)

      COMMON/COORD/NA,NB,NC,ND,ILAZY,RCD,ZHBCD,PABCD

      COMMON/CORDU/T(90,90),NOAT,KWIK,R12,R23,THETA

      ENTRY OUT1

      WRITE(108,959)

959   FORMAT(1H1)

      WRITE(108,900)(NAME(I),I=1,20)

900   FORMAT(20A4)

      WRITE(108,951)NOAT,(IZAT(I),I=1,3),KWIK,R12,R23,THETA

951   FORMAT(5X,'NOAT =',I2,3X,'IZAT(1) =',I2,2X,'IZAT(2) =',I2,

     C                      2X,'IZAT(3) =',I2,5X,'KWIK =',I1,3X,

     C                            'R12 =',F7.4,2X,'R23 =',F7.4,3X

     C                            'THETA =',F14.7)

      RETURN

      ENTRY OUT2

      WRITE(108,953)

953   FORMAT(20X,'NA',3X,'NB',3X,'NC',3X,'ND',5X,'IZAT(ND)',5X,

     C            'ILAZY',8X,'RCD',12X,'THBCD',15X,'PABCD'/)

      RETURN

      ENTRY OUT3

      WRITE(108,954)NA,NB,NC,ND,IZAT(ND),ILAZY,RCD,THBCD,PABCD

954   FORMAT(20X,I2,3(3X,I2,8X,I2,I0X,I1,7X,F7.4,2(5X,F14.7)/)

      RETURN

      ENTRY OUT4

      WRITE(108,959)
```

```
      WRITE(108,900)(NAME(I),I=1,20)

      WRITE(108,955)

      WRITE(108,956)(I,X(I),Y(I),Z(I),I=1,NOAT)

      WRITE(108,959)

  955 FORMAT(7X,'NO.OF ATOM',10X,'X-COORDINATE"',15X,

    C            'Y-COORDINATE',15X,2-COORDINATE'/)

  956 FORMAT(12X,I2,15X,F10.7,17X,F10.7,17X,F10.7)

      RETURN

      ENTRY OUT5

      WRITE(108,900)(NAME(I),I=1,20)

      WRITE(108,957)

  957 FORMAT(1H0,21HINTERATOMIC DISTANCES,//)

      RETURN

      END

      SUBROUTINE BUFOUTMT

      INTEGER SW

      COMMON/CURENT/SW,NATC,IZAT(50),X(50),Y(50),Z(50)

      COMMON/STANDARD/MS,NATS,IZAS(30),XS(30),YS(30),ZS(30)

      CALL BUFFER OUT(1,1,SW,122,II,III)

      IF(II.EQ.4)WRITE(108,900)SW

      IF(SW.EQ.MS)GO TO 10

      RETURN

   10 NATS=NATC

      DO 20 I=1,NATC

      IZAS(I)=IZAT(I)

      XS(I)=X(I)
```

```
      YS(I)=Y(I)

  20  ZS(I)=Z(I)

      RETURN

 900  FORMAT(///10X,'**ERROR**  BUFFER OUT MOLECULE  ',I4)

      END

      SUBROUTINE MSD

      COMMON/CURENT/SW,NATC,IZAT(50),X(50),Y(50),Z(50)

      COMMON/STANDARD/MS,NATS,IZAS(30),XS(30),YS(30),ZS(30)

      COMMON VOL,NI,NT

      COMMON/RANDOM/I1,I2,S,J1,J2,P,K1,K2,Q

      COMMON/LIMIT/XMAXP, YMAXP,ZMAXP,XMAXN,YMAXN,ZMAXN

      COMMON/VANDER/WAALS(100),NRWW

      NI=0

      NT=0

      I1=65539

      N=MAX0(NATS,NATC)

      DO 40 J=1,10000

      CALL UNIREL(XMAXN,XMAXP,S,I1,XX)

      CALL UNIREL(YMAXN,YMAXP,P,I1,XX)

      CALL UNIREL(ZMAXN,ZMAXP,Q,I1,XX)

      MP=0

      MC=0

      DO 20 I=1,N

      IF(I.GT.NATS)GO TO 10

      IF(((S-XS(I))**2+(P-YS(I))**2+(Q-ZS(I))**2).LT.

    C     WALLS(IZAS(I))**2)MP=MP+1
```

```
10 IF(I.GT.NATC)GO TO 15

   IF(((S-X(I))**2+(P-Y(I))**2+(Q-Z(I))**2).LT.

 C     WALLS(IZAT(I)**2)MC=MC+1

15 IF(MP.GE.1.AND.MC.GE.1)GO TO 30

20 CONTINUE

   IF(MP.EQ.0.AND.MC.EQ.0)GO TO 30

   NI=NI+1

30 NT=NT+1

40 CONTINUE

   CALL OUTMSD

   RETURN

   END

   SUBROUTINE OUTMSD

   INTEGER SW

   COMMON/CURENT/SW,NATC,IZAT(50),X(50),Y(50),Z(50),

   COMMON VOL,NI,NT,

   COMMON/TMSD/TMSD(100)

   IF(SW.EQ.1)WRITE(108,900)

   TMSD(SW)=FLOAT(NI)/FLOAT(NT)*VOL

   WRITE(108,910)SW,TMSD(SW)

900 FORMAT(/////40X,'MOLECULE      MSD'/40X,16(1H*)///)

910 FORMAT(42X,I4,3X,F9.4)

   RETURN

   END

   SUBROUTINE BUFINPMT(*)

   INTEGER SW
```

```
      COMMON/CURENT/SW,NATC,IZAT(50),X(50),Y(50),Z(50)

      CALL BUFFER IN(1,1,SW,122,II,III)

      IF(II.EQ.3)GO TO 10

      IF(II.EQ.4)WRITE(108,900)SW

  900 FORMAT(///10X,'**ERROR**   BUFFER IN MOLECULE   ',I4)

      RETURN

   10 RETURN 1

      END

      SUBROUTINE MAX

      REAL MXSP,MYSP,MZSP,MYCN,MZCN

      COMMON VOL,NI,NT

      COMMON/STANDARD/MS,NATS,IZAS(30),XS(30),YS(30),ZS(30)

      COMMON/VANDER/WAALS(100),NRW

      COMMON/CURENT/SW,NATC,IZAT(50),XC(50),YC(50),ZC(50)

      COMMON/LIMIT/XMAXP,YMAXP,ZMAXP,XMAXN,YMAXN,ZMAXN

      MXSP=0.0

      MYSP=0.0

      MZSP=0.0

      MXSN=0.0

      MYSN=0.0

      MZSN=0.0

      MXCP=0.0

      MYCP=0.0

      MZCP=0.0

      MXCN=0.0

      MYCN=0.0
```

```
      MZCN=0.0

      WMAX=0.0

      DO 10 I=1,NATS

      IF(XS(I).GE.MXSP)MXSP=XS(I)

      IF(YS(I).GE.MYSP)MYSP=YS(I)

      IF(ZS(I).GE.MZSP)MZSP=ZS(I)

      IF(XS(I).LE.MXSN)MXSN=XS(I)

      IF(YS(I).LE.MYSN)MYSN=YS(I)

      IF(ZS(I).LE.MZSN)MZSN=ZS(I)

   10 CONTINUE

      DO 20 I=1,NATC

      IF(XC(I).GE.MXCP)MXCP=XC(I)

      IF(YC(I).GE.MYCP)MYCP=YC(I)

      IF(ZC(I).GE.MZCP)MZCP=ZC(I)

      IF(XC(I).LE.MXCN)MXCN=XC(I)

      IF(YC(I).LE.MYCN)MYCN=YC(I)

      IF(ZC(I).LE.MZCN)MZCN=ZC(I)

   20 CONTINUE

      DO 30 I=1,NRW

      IF(WAALS(I).GE.WMAX)WMAX=WAALS(I)

   30 CONTINUE

      XMAXP=AMAX1(MXSP,MXCP)

      YMAXP=AMAX1(MYSP,MYCP)

      ZMAXP=AMAX1(MZSP,MZCP)

      XMAXN=AMIN1(MXSN,MXCN)

      YMAXN=AMIN1(MYSN,MYCN)
```

```
      ZMAXN=AMIN1(MZSN,MZCN)

      XMAXP=XMAXP+WMAX

      XMAXN=XMAXN-WMAX

      YMAXP=YMAXP+WMAX

      YMAXN=YMAXN-WMAX

      ZMAXP=ZMAXP+WMAX

      ZMAXN=ZMAXN-WMAX

      VOL=(XMAXP+ABS(XMAXN))*(YMAXP+ABS(YMAXN))*(ZMAXP+ABS(ZMAXN))

      RETURN

      END

      SUBROUTINE MODIF

      COMMON/GRAD/M,N

      COMMON/TMSD/TMSD(100)

      COMMON/GAP/SIG(20,50),SIGB(20),SIGR(20,50)

      M=M+1

      DO 10 I=1,N

      SIG(M,I)=TMSD(I)

   10 CONTINUE

      RETURN

      END

      SUBROUTINE VANWAALS

      COMMON/VANDER/WAALS(100),NRW

      DO 1 I=1,100

    1 WAALS(I)=0.0

      WAALS(1)=1.2

      WAALS(6)=1.57
```

```
WAALS(7)=1.5

WAALS(8)=1.4

WAALS(16)=1.85

WAALS(15)=1.9

WAALS(43)=2.0

WAALS(9)=1.35

WAALS(17)=1.80

WAALS(35)=1.95

WAALS(53)=2.15

NRW=53

RETURN

END
```

REFERENCES

1. E.Fischer : Chem.Ber. 27, 298 S (1894)
2. W.E.Purcell,G.E.Bass,J.M.Clayton :"Strategy of Drug Design, A Guide to Biological Activity". New York; Wiley 1973
3. W.E.Purcell, In : "Biological Activity and Chemical Structure" pg.282. Jane A.Keverling-Buisman,(ed.),Elsevier Sci.Publ. 1977
4. A.Verloop,J.Tipker, ibid. p.63
5. C.Hansch, ibid p.47
6. R.Frankem, ibid p.251; a) R.Frankem "Optimisierungsmethoden in der Wirkstofforschung", Berlin: Akademie Verlag, 1979
7. R.W.Taft,Jr, In: "Steric Effects in Organic Chemistry". chap.3. M.S.Newman, (ed.). New York: Wiley 1956
8. J.E.Amoore, Ann.N.Y. Acad.Sci. 116, 457 (1964)
9. J.E.Amoore,G.Palmieri,E.Wanke, Nature 216, 1084 (1967)
10. J.E.Amoore, Nature 214, 1095 (1967)
11. N.L.Allinger : Proc.Int.Congr. Pharmacol. 5, 5763 (1973), R.H.Karger (ed.). Basel 1973
12. A.Verloop,W.Hoogenstraaten,J.Tipker, Drug Design 7, 165 (1976); E.J.Ariens, (ed.) New York: Academic Press 1976
13. A.T.Balaban : Theoret.Chim.Acta (Berlin) in press
14. Z.Simon,Z.Szabadai, Stud.Biophys.(Berlin) 39, 123 (1973)
15. Z.Simon, Angew.Chem. 86,802(1974); Internat.Ed. 13,719(1974), a) Z.Simon "Quantum Biochemistry and Specific Interactions" chap.6. Tunbridge Wells: Abacus Press 1976
16. Z.Simon, Stud.Biophys.(Berlin) 51, 49 (1975)
17. I.Motoc et al. : Preprint, Univ.Timisoara, Serie Chim.(1975)Nr.1
18. I.Motoc,S.Holban,D.Ciubotariu, ibid (1975) Nr.16
19. I.Motoc et al.: Stud.Biophys.(Berlin) 66, 75 (1977)
20. Z.Simon et al.: Stud.Biophys.(Berlin) 55, 217 (1976) : a) Z.Simon et al.: Preprint, Univ.Timisoara, Serie Chim.(1978), Nr.1; b) Z.Simon, et al.: In : "Symposium of Chemical Structure-Biological Activity Relationships·Quantitative Approaches" p.161-167. P.Oehme,R.D.Franke (eds.) Berlin: Akademie Verlag 1978; c) Z.Simon,I.I.Badilescu,T.Racovitan, MATCH 3,257 (1977)
21. Z.Simon,I.I.Badilescu,T.Racovitan: J.theor.Biol.66, 485 (1977)
22. R.S.Schaare,A.N.Martin, J.Pharm.Sci. 54, 1707 (1965)
23. A.Verloop, Symp.Royal Neth.Chem.Soc. Design Bioactive Compounds, pp.17-31, 1970

24. R.R.Sokol, Science, _185_, 1115 (1974)

25. C.Hansch,S.H.Unger,A.B.Forsythe, J.Med.Chem. _16_, 1217 (1973)

26. W.J.Dunn III, M.J.Greenberg and S.S.Callejas, J.Med.Chem., _19_, 1299 (1976)

27. W.Spendley,G.R.Hext,F.R.Himsworth, Technometrics _4_, 441 (1962)

28. F.Darvas, J.Med.Chem. _17_, 799 (1975)

29. H.G.Boxenbaum,S.Riegelman,R.M.Elashoff, J.Pharmacokin. Biopharm. _2_, 123 (1974)

30. D.M.Olsson,L.S.Nelsonm, Technometrics _17_, 45 (1975)

31. R.Franke, W.Meisske, Acta Biol.Med.Ger. _35_, 73 (1976)

32. R.Hagemann,R.Franke, Pharmazie _31_, 2 (1976)

33. C.E.Klopfenstein,C.L.Wilkis (eds.), "Computers in Chemical and Biochemical Research", vol.2, New York, Academic Press 1974

34. H.C.Andrews, "Introduction to Mathematical Techniques in Pattern Recognition". New York, Wiley 1972

35. P.C.Jurs, T.L.Isenhour, "Chemical Applications of Pattern Recognition". New York: Wiley 1975

36. A.Cammarata,G.K.Menon, J.Med.Chem. _19_, 739 (1976)

37. U.Grenander, "Pattern Analysis". Berlin: Springer 1978

38. W.P.Purcell,G.E.Bass,J.M.Clayton, "Strategy of Drug Design : A Guide to Biological Activity". New York: Wiley 1973

39. N.B.Chapman,J.Shorter (eds.). "Advances in Linear Free Energy Relationships". vol.1, London: Plenum 1972

40. N.B.Chapman,J.Shorter (eds.). "Advances in Linear Free Energy Relationships". vol.2, London: Plenum 1978

41. A.Verloop, In : "Drug Design", E.J.Ariens, (ed.) vol.3, New York: Academic Press 1972

42. N.R.Draper,H.Smith, "Applied Regression Analysis", New York : Wiley 1966

43. G.W.Snedecor, "Statistical Methods". 4th ed., Ames: Iowa State College Press 1946

44. O.L.Davies (ed.). "Statistical Methods in Research and Production". London: Oliver and Boyd 1961

45. G.W.Snedecor,W.C.Cochran, "Statistical Methods", 6th ed., Ames: Iowa State Univ.Press 1967

46. R.R.Hocking, Biometrics _32_, 1 (1976)

47. R.L.Mason, Comm.Statist. _4_, 277 (1975)

48. J.C.Topliss,R.J.Costello, J.Med.Chem. _15_, 1966 (1972)

49. C.Gini, Metron _1_, 63 (1921)

50. S.Wold, Chem.Scr. _5_, 97 (1974)

51. J.Hine, "Structural Effects on Equilibria in Organic Chemistry". New York: Wiley 1975

52. P.R.Wells, "Linear Free Energy Relationships". London : Academic Press 1968

53. J.F.Danelli,J.F.Moran,D.J.Triggle (eds.). "Fundamental Concepts in Drug-Receptor Interactions". New York: Academic Press 1970

54. R.F.Beers,Jr.,E.G.Bassett (eds.). "Cell Membrane Receptors for Viruses, Antigens and Antibodies, Polypeptide Hormones and Small Molecules". New York: Raven Press 1976

55. R.J.P.Williams, Angew.Chem.Int.Ed.Engl. 16,766 (1977)

56. W.G.Richards,R.Clarkson,C.R.Ganellin, Phil.Trans.Roy.Soc.(B) 272, 75 (1975)

57. J.C.McGowan, J.Appl.Chem. 2, 323 (1952); 4, 41 (1954)

58. J.C.McGowan, Nature 200, 1317 (1963)

59. O.Exner, Collect.Czech.Chem.Comm. 32, 1 (1967)

60. A.Bondi, J.Phys.Chem. 62, 441 (1964)

61. E.J.Lien, Am.J.Pharm.Educ. 33, 368 (1969)

62. R.A.Coburn,A.J.Solo, F.Med.Chem. 19, 748 (1976)

63. C.K.Ingold, J.Chem.Soc. 1930, 1032

64. J.Shorter, In : "Advances in Linear Free Energy Relationships". Vol.1,Ch.2, N.B.Chapman,J.Shorter (eds.). London: Plenum 1972

65. C.K.Hancock,E.A.Meyers,B.J.Yager, J.Amer.Chem.Soc. 83, 4211 (1961)

66. C.K.Hancock,C.P.Falls, J.Amer.Chem.Soc. 83, 4214 (1961)

67. C.K.Hancock, J.Org.Chem. 30, 1174 (1965)

68. C.K.Hancock, J.Org.Chem. 38, 4239 (1973)

69. V.A.Palm, "Fundamentals of the Quantitative Theory of Organic Reactions" Ch.10 (russ.). Leningrad: Khimija 1967

70. I.V.Tavlik,V.A.Palm, Org.React. (USSR), 2, 445 (1971)

71. S.H.Unger, Ph.D.Thesis.MIT, September, 1970

72. S.H.Unger,C.Hansch, Progr.Phys.Org.Chem. 12, 91 (1976)

73. C.G.Swain,E.C.Lupton, J.Amer.Chem.Soc. 90, 4328 (1968)

74. M.Charton, J.Am.Chem.Soc. 91, 615 (1969)

75. M.Charton, Progr.Phys.Org.Chem. 8, 235 (1971)

76. E.Kutter,C.Hansch, J.Med.Chem. 12, 647 (1969)

77. L.Cronnenberger,B.Dolfin,H.Pacheco, C.R.Acad.Sci. (Paris) D269, 1334 (1969)

78. I.Motoc, Analele Univ.Timisoara, Ser.Fiz Chim XII(1), 57 (1974)

79. M.Charton : personal communication

80. L.Ruzicka, Chem.Ztg. 44, 129 (1920)

81. J.E.Amoore,D.Venstrom, Proc. 2nd Intern.Symp.Olfaction and
     Taste, Tokyo 1965

82. I.Motoc, unpublished

83. K.Wilson, 144th Nat.Amer.Chem.Soc.Mat., Spring 1963, Div.Chem.
     Ed. 23

84. J.J.Kaufman,D.Estok,K.Feldmann, unpublished, cited in ref.77

85. C.Hansch,D.F.Calef, J.Org.Chem. $\underline{41}$, 124 (1976)

86. C.Grieco, et al.: J.Med.Chem. $\underline{20}$, 586 (1977)

87. C.Hansch,J.Y.Fukunaga,P.Y.C.Jow, J.Med.Chem. $\underline{20}$, 96 (1977)

88. C.Hansch, et al.: J.Med.Chem. $\underline{16}$, 1207 (1973)

89. C.Hansch, J.Med.Chem. $\underline{20}$, 304 (1977)

90. A.Leo, et al.: J.Med.Chem. $\underline{18}$, 865 (1975)

91. J.J.Kaufman, Int.J.Quantum Chem. QBS, $\underline{4}$, 375 (1977)

92. a) I.Motoc,F.Kerek, to be published; b) O.Dragomir,
     I.Muscutariu,I.Motoc, to be published

93. R.J.Boyd, J.Phys. B $\underline{10}$, 2283 (1977)

94. J.L.Webb, "Enzymes and Metabolic Inhibitors". New York: Academic
     Press 1963

95. I.Motoc,O.Dragomir, to be published

96. C.Hansch,P.Moser, personal communication

97. The documentation and the listing of the STERIMOL program were
     kindly supplied by Dr.Hoogenstraaten

98. C.Hansch, Drug Metab.Rev. $\underline{1}$, 1 (1972)

99. A.Verloop,C.D.Ferrell, in "Pesticide Chemistry in the 20th
     Century", J.R.Plimmer (ed.). ACS Symposium Series, vol.37,
     pp.237-270, ACS, Washington 1977

100. T.C.Bruice,N.Kharasch,R.J.Winzler, Arch.Biochem.Biophys. $\underline{62}$,
     305 (1956)

101. S.M.Free,Jr.,J.W.Wilson, J.Med.Chem. $\underline{7}$, 395 (1964)

102. C.Hansch,C.Silipo,E.E.Steller, J.Pharm.Sci. $\underline{64}$, 1186 (1975)

103. C.Hansch,D.F.Calef, J.Org.Chem. $\underline{41}$, 1240 (1976)

104. C.Hansch,M.Yoshimoto,M.H.Doll, J.Med.Chem. $\underline{19}$, 1089 (1976)

105. C.Hansch, et al.: Arch.Biochem.Biophys. $\underline{183}$, 383 (1977)

106. L.P.Hammett, "Physical Organic Chemistry". 2nd ed., New York:
     McGraw Hill 1970

107. H.C.Brown, Adv.Phys.Org.Chem. $\underline{1}$, 35 (1963)

108. H.H.Jaffé, Chem.Rev. $\underline{53}$, 191 (1953)

109. H.van Bekkum,P.E.Verkade,B.M.Wepster, Rec.Trav.Chim.Pays-Bas.
     $\underline{78}$, 815 (1959)

110. R.W.Taft, J.Phys.Chem. $\underline{64}$, 1805 (1960)

111. J.Shorter, "Correlation Analysis in Organic Chemistry", pp.36.
     Oxford: Clarendon 1973

112. G.G.Smith,G.O.Larson, J.Amer.Chem.Soc. 82, 99 (1960)

113. R.W.Taft, J.Am.Chem.Soc. 75, 4231 (1953)

114. J.Shorter, "Correlation Analysis in Organic Chemistry". pp.40:
     Oxford: Clarendon 1973

115. T.A.Mastryukova,M.I.Kabachnik, Uspekhi khim. 38, 1751 (1969)

116. F.Peradejordi,A.N.Martin,A.Cammarata, J.Pharm.Sci. 60, 576
     (1971)

117. R.L.Schaare, In: "Drug Design" vol.1, pp.405-449. E.J.Ariens
     (ed.). London: Academic Press 1971

118. L.B.Kier, "Molecular Orbital Theory in Drug Research".
     New York: Academic Press 1971

119. J.J.Kaufman, Inter.J.Quantum Chem. QBS,4,375 (1977), p.405

120. M.Charton, J.Org.Chem. 30, 3341 (1965)

121. W.Kauzmann, Advan.Protein Chem. 14, 1 (1959)

122. G.Nemethy,H.A.Scheraga, J.Chem.Phys. 36, 3401 (1962)

123. G.Nemethy,H.A.Scheraga, J.Chem.Phys. 36, 3382 (1962)

124. I.M.Klotz, J.Amer.Chem.Soc. 81, 5519 (1959)

125. I.M.Klotz, Brookhaven Symp.Biol. 13, 25 (1960)

126. C.Tanford, Science 200, 1012 (1978)

127. A.Suggett, In: "Biological Activity and Chemical Structure".
     pp.95. J.A.K.Buisman (ed.). Amsterdam: Elsevier 1977

128. R.F.Rekker, ibid., pp.107

129. C.Hansch, et al.: Nature 194, 180 (1962)

130. C.Hansch,M.Streich, J.Amer.Chem.Soc. 85, 2817 (1963)

131. C.Hansch,T.Fujita, J.Amer.Chem.Soc. 86, 1616 (1964)

132. A.Leo,C.Hansch,D.Elkins, Chem.Rev. 71, 525 (1971)

133. C.Hansch,W.J.Dunn III, J.Pharm.Sci. 61, 1 (1972)

134. S.S.Davis, J.Pharm.Pharmacol. 25, 193 (1973)

135. I.Moriguchi, Chem.Pharm.Bull. 23, 247 (1975)

136. R.F.Rekker, "The Hydrophobic Fragmental Constant, Its Deriva-
     tion and Application. A means of Characterizing Membrane Sys-
     tems". Amsterdam: Elsevier 1977

137. Y.C.Martin, In: "Drug Design" p. 100, E.J.Ariens (ed.).
     New York: Academic Press vol.8 (1977)

138. Z.Simon, Rev.Roum.Biochim. 5, 319 (1968)

139. Z.Szabadai,Z.Simon, Rev.Roum.Biochim. 9, 327 (1972)

140. Z.Simon,E.Draskovits-Schuch, Rev.Roum.Chim. 22, 87 (1977)

141. P.Seiler, Chim.Therap. 9, 663 (1974)

142. I.Moriguchi,Y.Kanada,K.Komatsu, Chem.Pharm.Bull. $\underline{24}$,1789 (1976)

143. B.Hetnarski,R.D.O'Brien, J.Med.Chem. $\underline{18}$, 29 (1975)

144. F.Harary, "Graph Theory". Reading, Mass.: Addison-Wesley 1969

145. R.J.Wilson, "Introduction to Graph Theory". Edinburgh: Oliver and Boyd 1972

146. A.T.Balaban (ed.). "Chemical Application of Graph Theory". London: Academic Press 1972

147. D.H.Rouvray,A.T.Balaban, In: "Applications of Graph Theory". R.J.Wilson,L.W.Beineke, (eds.). London: Academic Press 1979 , D.H.Rouvray,R.I.C.Rev. $\underline{4}$, 173 (1971); Chem.Brit. $\underline{10}$, 11 (1974); J.Chem.Educ. $\underline{52}$, 768 (1975)

148. G.M.Dyson, "A New Notation and Enumeration System for Organic Compounds". 2nd Ed., London: Longmans 1949; "Rules for I.U.P.A.C. Notation for Organic Compounds". London: Longmans 1961

149. W.J.Wiswesser, "A Line-Formula Chemical Notation". New York: T.Y.Crowell Co. 1954

150. J.Lederberg, et al.: J.Am.Chem.Soc. $\underline{91}$, 2973 (1969)

151. R.C.Read,R.S.Miller, "A New System for the Designation of Chemical Compounds for the Purpose of Data Retrieval. I. Acyclic Compounds", Preprint, Jamaica: Univ.of the West Indies 1969

152. K.Ruedenberg, J.Chem.Phys. $\underline{22}$, 1878 (1954); D.H.Rouvray, chapter 7 in ref. 146 above; A.Graovac,I.Gutman,N.Trinajstić, "Topological Approach to the Chemistry of Conjugated Molecules". Lecture Notes in Chemistry, $\underline{4}$ (1976)

153. L.Spialter, J.Chem.Doc. $\underline{4}$, 261, 259 (1964); J.Am.Chem.Soc. $\underline{85}$, 2012 (1963)

154. A.T.Balaban,F.Harary, J.Chem.Doc. $\underline{11}$, 258 (1971)

155. Y.Kudo,T.Yamasaki,S.I.Sasaki, J.Chem.Doc. $\underline{13}$, 225 (1973)

156. W.C.Hernson, J.Chem.Doc. $\underline{14}$, 150 (1974), cf.also M.Randić, N.Trinajstić, T.Zivković, J.Chem.Soc. Faraday Trans. II, $\underline{72}$, 244 (1976); T.Zivković,N.Trinajstić,M.Randić, Mol.Phys. $\underline{30}$, 517 (1975); M.Randić, in press

157. M.Randić, J.Chem.Phys. $\underline{60}$, 3920 (1974)

158. M.Randić, J.Chem.Inf. Computer Sci. $\underline{2}$, 105 (1975); Chem.Phys. Lett. $\underline{42}$, 283 (1976); J.Chem.Phys. $\underline{62}$, 309 (1975)

159. A.L.Mackay, J.Chem.Phys. $\underline{62}$, 308 (1975)

160. D.H.Rouvray, Amer.Sci. $\underline{61}$, 729 (1973); MATCH, $\underline{1}$, 125 (1975)

161.   a) H.Wiener,   J.Am.Chem.Soc. <u>69</u>, 17 (1947);   b) ibid. 2636;
       c) ibid., J.Chem.Phys. <u>15</u>, 766 (1947);   d) ibid. J.Phys.Chem.
       <u>52</u>, 425 (1948);   e) ibid. 1082

162.   J.R.Platt, J.Chem.Phys. <u>15</u>, 419 (1947)

163.   D.H.Rouvray, B.C.Crafford, S.Afr.J.Sci. <u>72</u>, 47 (1976)

164.   K.Altenburg, Kolloid-Ztschr. 178, 112 (1971); Brennst.Chem.
       <u>47</u>, 100, 331 (1966)

165.   a) H.Hosoya, Bull.Chem.Soc.Jpn. <u>44</u>, 2332 (1971); b) Internat.
       J.Quantum Chem. <u>6</u>, 801 (1972);   c) J.Chem.Doc. <u>12</u>, 181 (1972);
       d) Chem.Lett. (Japan), 65 (1972);   e) Fibonacci Quart. <u>3</u>, 255
       (1973)

166.   J.R.Platt, J.Phys.Chem. <u>56</u>, 328 (1952)

167.   M.Gordon, G.R.Scantlebury, Trans. Faraday Soc. <u>60</u>, 605 (1964)

168.   H.Hosoya, K.Kawasaki, K.Mizutani, Bull.Chem.Soc.Jpn. <u>45</u>, 3415
       (1972)

169.   K.Mizutani, K.Kawasaki, H.Hosoya, Nat.Sci.Rep. Ochanomizu Univ.
       <u>22</u>, 39 (1971)

170.   K.Kawasaki, K.Mizutani, H.Hosoya, ibid. <u>22</u>, 181 (1971)

171.   H.Hosoya,M.Murakami,M.Gotoh, ibid. <u>24</u>, 27 (1973)

172.   I.Gutman et al.: J.Chem.Phys. <u>62</u>, 3399 (1975)

173.   M.Randić, J.Am.Chem.Soc. <u>97</u>, 6609 (1975)

174.   L.B.Kier et al.: J.Pharm.Sci. <u>65</u>, 1226 (1976)

175.   L.B.Kier et al.: J.Pharm.Sci. <u>64</u>, 1971 (1975)

176.   L.Lovasz, J.Pelikan, Period.Math.Hung. <u>3</u>, 175 (1973)

177.   D.Bonchev,N.Trinajstić, J.Chem.Phys. <u>67</u>, 4517 (1977)

178.   D.Bonchev,A.T.Balaban,O.Mekenyan, to be published

179.   E.A.Smolenskii, Zhur.fiz.khim. <u>38</u>, 1288 (1964)

180.   H.J.Bernstein, J.Chem.Phys. <u>19</u>, 140 (1951); <u>20</u>, 263, 351 (1952);
       J.Phys.Chem. <u>69</u>, 1550 (1965)

181.   V.M.Tatevskii, "Khimicheskoe stroenye uglevodorodov i zakono-
       mernosti v ikh fiziko-khimicheskikh svoistv", Moscow 1953;
       V.M.Tatevskii,V.A.Benderskii,S.S.Yarovoi, "Metody rascheta
       fiziko-khimicheskikh svoistv parafinovykh uglevodorodov", Moscow
       1960

182.   V.M.Tatevskii,Yu.G.Papulov, Zh.Obshchei Khim. <u>34</u>, 241, 489, 708
       (1960); Yu.G.Papulov et al.: Zh.fiz.khim. <u>48</u>, 31 (1974)

183.   R.F.Muirhead, Proc.Edinburgh Math.Soc. <u>21</u>, 144 (1903); <u>19</u>, 36
       (1901), <u>24</u>, 45 (1906); cf. also E.Ruch,A.Schönhofer, Theor.
       Chim.Acta <u>19</u>, 225 (1970); E.Ruch, ibid. <u>38</u>, 167 (1975);
       E.Ruch,A.Mead, ibid. <u>41</u>, 95 (1976)

184. I.Gutman,M.Randić, Chem.Phys.Lett. 47, 15 (1977)

185. W.J.Taylor,J.M.Rignocco,F.D.Rossini, J.Res.Natl.Bur.Standards, 34, 413 (1945)

186. A.T.Balaban, I.Motoc, Math.Chem. 5, 197 (1979)

187. M.Randić, Chem.Phys.Letters, 53, 602 (1978)

188. K.Burnham et al.: J.Am.Chem.Soc. 99, 1836 (1977)

189. H.F.Hameka, J.Chem.Phys. 34, 1966 (1961)

190. L.B.Kier,L.H.Hall, "Molecular Connectivity in Chemistry and Drug Research". New York: Academic Press 1976

191. E.Kováts, Z.analyt.Chem. 181, 351 (1961)

192. D.Bonchev,N.Trinajstić, J.Thermodyn. (submitted); D.Bonchev, N.Trinajstić, Internat.J.Quantum Chem. 12, 293 (1978); O.Mekenyan,G.Protić,N.Trinajstić, J.Chromatogr. (in press)

193. L.B.Kier, J.Pharm.Sci. 61, 1394 (1972)

194. H.D.Holtje,L.B.Kier, J.Pharm.Sci. 63,1435 (1974); J.Med.Chem. 17, 814 (1974), J.Theor.Biol. 48, 197 (1974)

195. D.Agin,L.Hersh,D.Holtzman, Proc.Natl.Acad.Sci. U.S.A. 53, 952 (1965)

196. L.B.Kier et al.: J.Pharm.Sci. 64, 1971 (1975)

197. R.B.Hermann, J.Phys.Chem. 76, 2754 (1972)

198. W.J.Murray,L.H.Hall,L.B.Kier, J.Pharm.Sci. 64, 1978 (1975)

199. L.B.Kier,W.J.Murray,L.H.Hall, J.Med.Chem. 18, 1272 (1975)

200. W.J.Murray,L.B.Kier,L.H.Hall, J.Mod.Chem. 19, 573 (1976)

201. C.Hansch, In: "Medicinal Chemistry". E.J.Ariens (ed.). vol.1, New York: Academic Press; C.Hansch,J.E.Quinlan,G.L.Lawrence, J.Org.Chem. 33, 347 (1968)

202. A.Leo,C.Hansch,D.Elkins, Chem.Rev. 71, 525 (1971)

203. G.G.Nys,R.F.Rekker, Chem.Ther. 5, 271 (1973)

204. C.Hansch,W.J.Dunn III, J.Pharm.Sci. 61, 1 (1972)

205. D.J.Crisp,D.H.A.Marr, Proc.2nd Int.Congr.Surface Activity, 310 (1957)

206. H.M.Vernon, J.Physiol. 47, 15 (1913)

207. E.Overton, "Studies on Narcosis". Jena: Fischer 1901

208. F.Batelli,L.Stern, Biochem.Z. 52, 226 (1913)

209. B.R.Baker,M.Kawaza, J.Med.Chem. 10, 302 (1967)

210. H.J.Schaeffer et al.: J.Med.Chem. 13, 452 (1970)

211. J.M.Clayton,W.P.Purcell, J.Med.Chem. 13, 452 (1970)

212. C.Hansch,J.M.Clayton, J.Pharm.Sci. 62, 1 (1973)

213. E.J.Lien, In "Medicinal Chemistry". J.Maas (ed.). vol.4, Amsterdam: Elsevier 1974

214. a. L.H.Hall,L.B.Kier,W.J.Murray, J.Pharm.Sci. 64, 1974 (1975);

b. L.B.Kier,L.H.Hall, ibid. <u>65</u>, 1806 (1976)

215. Z.Simon et al.: to be published

216. N.Fujino et al.: J.Med.Chem. <u>16</u>, 1144 (1973)

217. A.Chiriac,Z.Simon,R.Vilceanu, Reprint, Univ.Timisoara, Ser.Chim. (1974) Nr.4

218. M.S.Kabachnik,M.P.Brestkov,M.Ya.Michelson , "IX Mendeleevskii Siezd." p.231 Izd.Nauka, Moscow 1963

219. Z.Simon,I.Motoc,S.Holban,D.Ciubotariu, Nr.1. Preprint, Univ. Timisoara, Ser.Fiz.Chim. (1975)

220. R.Vilceanu et al.: Analele Univ.Timisoara, Serie Fiz.Chim. XV(2), 43 (1977) Publ 1978

221. A.Chiriac,Z.Simon,R.Vilceanu, Studia Biophys.(Berlin) <u>51</u>, 183 (1975)

222. E.Draskovits-Schuch, Z.Simonet al.: to be published

223. J.Miklos,Z.Simon, Rev.Roum.Biochim. <u>13</u>, 283 (1976)

224. Z.Simon, Preprint, Univ.Timisoara Ser.Chim. (1974) Nr.5

225. R.D.O'Brien, "Insecticide Action and Metabolism". New York, London: Academic Press (1967)

226. R.L.Metcalf,T.R.Fukuto, J.Agric.Food.Chem. <u>13</u>, 220 (1965), <u>15</u>, 1022 (1967)

227. A.Chiriac et al.: Rev.Roum.Biochim. <u>12</u>, 143 (1975)

228. G.Schrader, "Die Entwicklung neuer Insektizide der Phosphor-säure Ester" . Verlag Chemie,GMBH,Weinheim (1963)

229. R.Vilceanu et al.: Studia Biophys. (Berlin) <u>34</u>, 1 (1972)

230. R.Vilceanu et al.: Rev.Roum.Biochim. <u>10</u>, 239 (1973)

231. A.Chiriac, Veronica Chiriac,Z.Simon, Analele Univ.Timisoara, Ser.Fiz.Chim. <u>XIV</u>(2), 53 (1976), Publ 1978

232. A.Chiriac et al.: Preprint Univ.Timisoara (1976) Nr.2,3,4,5

233. T.A.Connors et al.: Chem.Biol.Interactions <u>5</u>, 415 (1977)

234. Z.Simon et al.: Rev.Roum.Biochim. <u>14</u>, 117 (1977)

235. I.Motoc et al.: Canad.J.Pharmacol. in press.

236. M.Charton, Progr.Phys.Org.Chem. <u>10</u>, 81 (1973)

237. Z.Simon, Stud.Biophys.(Berlin) <u>62</u>, 167 (1977)

238. I.I.Badilescu,Z.Simon, Rev.Roum.Biochim <u>13</u>, 239 (1976)

239. G.Hein,C.Niemann, Proc.Natn.Ac.Sci.US <u>55</u>, 64 (1966)

240. S.Holban et al.: Rev.Roum.Biochim <u>16</u>, 99 (1979);
a) Z.Simon et al.: Stud.Biophys.(Berlin) <u>59</u>, 181 (1976)

241. J.R.Knowles, J.theor.Biol. <u>9</u>, 213 (1965)

242. M.S.Silver et al.: J.Amer.Chem.Soc. <u>92</u>, 315 (1970)

243. S.Holban,Z.Simon, Preprint Univ.Timisoara, Ser.Chim (1977) Nr.7

244. A.Dupaix,J.J.Bechet,C.Roucous, Biochemistry 12, 2559 2566 (1973)

245. C.Hansch et al.: J.Med.Chem. 20, 1420 (1977)

246. C.Grieco et al.: J.Med.Chem. 20, 586 (1977)

247. Ch.W.Shoppee, "Chemistry of Steroids". Chap.IV,TAB.XXVI
     London: Butterworth 1964

248. A.Vlad,Z.Simon: in preparation

249. A.Rotman et al.: J.Med.Chem. 18, 138 (1975)

250. Z.Simon,S.Holban, Analele Univ.Timisoara, Ser.Fiz. 12 (1974),
     63 (1978)

251. L.L.Iversen,B.Callingham, in "Fundamentals of Biochemical
     Pharmacology". Z.M.Bacq, (ed.). Oxford: Pergamon Press
     (1971), 253

252. M.G.Plauchitin,A.Vlad,Z.Simon, Rev.Roum.Biochimie 15,229 (1978)

253. M.C.Malcolm-Dyson,P.May, "May's Chemistry of Synthetic Drugs".
     Chap.22. London: Longmans (1955)

254. V.Popoviciu et al.: Studia Biophys. (Berlin) 69, 75 (1978)

255. I.M.Nazarov,L.S.Bergelson, "Himiia steroidnyh hormonov".
     Chap.VI. Izd.Akad.Nauk SSSR, Moscow (1955)

256. C.Silipo,C.Hansch, J.Amer.Chem.Soc. 97, 6849 (1975)

257. A.Stepan,I.I.Badilescu,Z.Simon, Analele Univ.Timisoara, Ser.Fiz.
     Chim. XV(2), 61 (1977) Publ.1979

258. C.Hansch,P.Moser, Immunochem. 15, 535 (1978)

259. I.I.Badilescu et al.: Analele Univ.Timisoara, Ser.Fiz.Chim.
     XIV(2), 29 (1976) Publ.1978

260. Z.Simon,S.Holban,I.Motoc, Rev.Roum.Biochim. 16, 141 (1979)

261. R.Lumry,E.L.Smith,J.Polglase, J.Phys.Chem. 55, 125 (1951)

262. K.Hahn, "Applications of Monte Carlo". RM-1237-AEC, Rand Corp.,
     1956

263. O.Taussky,J.Todd, "Generation and Testing of Pseudo-Random
     Numbers". New York: Wiley 1956

264. S.M.Ermakov, "Monte Carlo Method and Related Problems" (russ.),
     Nauka, Moscow, 1971

265. J.L.Webb, "Enzymes and Metabolic Inhibitors". New York: Academic
     Press 1963

266. R.J.Boyd, J.Phys. B, 10, 2283 (1977)

267. COORD, Program QCPE No.136

268. F.Borek, in "Immunogenicity, Physico chemical and Biological
     Aspects". F.Borek (ed.). Amsterdam:North Holland (1972)

269. S.Sella "Immunology, Immunopathology and Immunity". Chap.1
     2nd ed., New York: Harper and Row (1975)

270. D.Pressman,A.L.Grossberg "The Structural Basis of Antibody
     Specificity". p.60. New York: Benjamins

271. I.Motoc,R.Vancea,Z.Simon in preparation

272. I.Motoc, Math.Chem. 4, 113 (1978)

273. I.Motoc,Z.Simon, Preprint, Timisoara Univ. Serie Chim. (1975)
     Nr.15   .

274. N.Jardine,R.Sibson, "Mathematical Taxonomy". New York: Wiley
     1971

275. W.A.Beyer et al.: Los Alamos Scientific Laboratory Report,
     LA-4973 1972

276. W.A.Beyer et al.: Math.Biosci. 19, 9 (1974)

277. M.S.Waterman,T.F.Smith,W.A.Beyer, Adv.Math. 20, 367 (1976)

278. S.M.Ulam, Annu.Rev.Biophys.Bioeng. 1, 277 (1972)

279. P.H.Sellers, Siam J. Appl.Math. 26, 787 (1974)

280. T.H.Jukes, "Molecules and Evolution". New York: Columbia Univ.
     Press 1966

281. W.M.Fitch and E.Margoliash, Science 155, 279 (1976)

282. W.M.Fitch, J.Mol.Biol. 16, 9 (1966)

283. N.Dourbaki, "Eléments de Mathématique. Topologie Générale".
     Hermann, Paris, 1964

284. I.Motoc, Math.Chem. 5, 275 (1979)

285. Y.Suezsaki,N.Gor, Int.J.Pept.Protein Res. 7, 333 (1975)

286. J.J.Ray, QCPE Program No.226.

Note : MATCH or Math.Chem.is a journal edited by A.T.Balaban,
       A.S.Dreiding, A.Kerber and O.E.Polansky.

Reactivity and Structure

Concepts in Organic Chemistry

Editors: K. Hafner, J.-M. Lehn, C. W. Rees,
P. v. Rague Schleyer, B. M. Trost,
R. Zahradnik

This series will not only deal with problems of the reactivity and structure of organic compounds but also consider synthetical-preparative aspects.
Suggestions as to topics will always be welcome.

Volume 1: J. Tsuji
Organic Synthesis
by Means of Transition Metal Complexes
A Systematic Approach
1975. 4 tables. IX, 199 pages
ISBN 3-540-07227-6

Volume 2: K. Fukui
Theory of Orientation and Stereoselection
1975. 72 figures, 2 tables. VII, 134 pages
ISBN 3-540-07426-0

Volume 3: H. Kwart, K. King
d-Orbitals in the Chemistry of Silicon, Phosphorus and Sulfur
1977. 4 figures, 10 tables. VIII, 220 pages
ISBN 3-540-07953-X

Volume 4: W. P. Weber, G. W. Gokel
Phase Transfer Catalysis in Organic Synthesis
1977. 100 tables. XV, 280 pages
ISBN 3-540-08377-4

Volume 5: N. D. Epiotis
Theory of Organic Reactions
1978. 69 figures, 47 tables. XIV, 290 pages
ISBN 3-540-08551-3

Volume 6: M. L. Bender, M. Komiyama
Cyclodextrin Chemistry
1978. 14 figures, 37 tables. X, 96 pages
ISBN 3-540-08577-7

Volume 7: D. I. Davies, M. J. Parrott
Free Radicals in Organic Synthesis
1978. 1 figure. XII, 169 pages
ISBN 3-540-08723-0

Volume 8: C. Birr
Aspects of the Merrifield Peptide Synthesis
1978. 62 figures, 6 tables. VIII, 102 pages
ISBN 3-540-08872-5

Volume 9: J. R. Blackborow, D. Young
Metal Vapour Synthesis in Organometallic Chemistry
1979. 36 figures, 32 tables. XIII, 202 pages
ISBN 3-540-09330-3

Volume 10: J. Tsuji
Organic Synthesis with Palladium Compounds
1980. XII, 207 pages
ISBN 3-540-09767-8

Volume 11:
New Syntheses with Carbon Monoxide
Editor: J. Falbe
1980. 118 figures, 127 tables.
Approx. 490 pages
ISBN 3-540-09674-4

Volume 12: J. Fabian, H. Hartmann
Light Absorption of Organic Colorants
Theoretical Treatment and Empirical Rules
1980. 76 figures, 68 tables. Approx. 260 pages
ISBN 3-540-09914-X

Springer-Verlag
Berlin Heidelberg New York

Inorganic Chemistry Concepts

Editors:
M. Becke
C. K. Jørgensen
M. F. Lappert
S. J. Lippard
J. L. Margrave
K. Niedenzu
R. W. Parry
H. Yamatera

Springer-Verlag
Berlin
Heidelberg
New York

Lecture Notes in Chemistry